Mechanics and Adaptronics

Series Editors

Ziel der Buchreihe ist die Publikation neuer Ansätze in der Mechanik und Adaptronik. Diese Ansätze umfassen zum einen experimentelle Themen als auch solche, die numerischen Charakters sind und erstrecken sich über diverse Größenskalen. Experimentelle, zukunftsweisen Themen beinhalten Untersuchungen auf Mikro- bis hin zur Makroebene und berücksichtigen unterschiedlichste Felder, von der Mechanik über Temperatur bis hin zu unterschiedlichen chemischen Feldern. Auch auf Seiten der numerischen Simulation werden auf Grundlage der experimentellen Untersuchungen neue multiskalige Mehrfeldmodelle entwickelt, kalibriert und validiert. Eine Besonderheit stellt hier die Thematik der Adaptronik dar, die als Ziel die Integration neuer Funktionen in tragende Bauteile durch die Kombination konventioneller Konstruktionswerkstoffe mit aktiven Werkstoffsystemen verfolgt. So werden Funktionswerkstoffe betrachtet, die Energiewandler sind und die die klassischen lasttragenden und formgebenden Funktionen um sensorische und aktorische Funktionen erweitern. Aktorische und sensorische Eigenschaften ermöglichen in Kombination mit adaptiven, schnellen Reglern eine optimale Selbstanpassung an die jeweilige Einsatzumgebung. Durch die Zusammenführung von numerischem und experimentellem Know-how auf mechanische als auch adaptronischer Seite ist es möglich, Erkenntnisse aus der Natur im Sinne des bionischen Gedankens in die Technik zu übertragen und so neue wissenschaftliche Richtungen aufzuschlagen.

Angelina Eret

Dynamic and Fatigue Assessment of Heavy-Duty Engine Valves

Springer

Angelina Eret
Hamburg, Germany

ISSN 2731-6211 ISSN 2731-622X (electronic)
Mechanics and Adaptronics
ISBN 978-3-031-49150-4 ISBN 978-3-031-49151-1 (eBook)
https://doi.org/10.1007/978-3-031-49151-1

This Springer imprint is published by the registered company Springer Nature Switzerland AG
The registered company address is: Gewerbestrasse 11, 6330 Cham, Switzerland

Paper in this product is recyclable.

Dynamic and Fatigue Assessment of Heavy-Duty Engine Valves

Von der Fakultät für Maschinenbau
der Technischen Universität Carolo-Wilhelmina zu Braunschweig

zur Erlangung der Würde

einer Doktor-Ingenieurin (Dr.-Ing.)

genehmigte Dissertation

von:	Angelina Eret M.Sc.
geboren in:	Oranienburg

eingereicht am:	28.03.2023
mündliche Prüfung am:	25.08.2023
Vorsitz:	Prof. Dr. habil. Rainer Krull
Gutachter:	Prof. Dr.-Ing. Markus Böl
	Prof. Dr.-Ing. Peter Eilts

Acknowledgements

This thesis was developed during my time as a Ph.D. student at MAN Energy Solutions SE in Augsburg at the Department of Mechanics and Validation and the Institute of Mechanics and Adaptronics, Faculty of Mechanical Engineering at Braunschweig University of Technology.

First, I would like to thank my Ph.D. supervisor Prof. Markus Böl for his guidance and support of this research. I also would like to thank Prof. Peter Eilts for kindly accepting the co-examination of this thesis and Prof. Rainer Krull for taking over the chair of the Ph.D. defence.

I would like to thank all of my former co-workers at the Department of Mechanics and Validation at MAN Energy Solutions SE for their kind support, fruitful discussions and an enjoyable working environment.

Especially I would like to thank Peter Böhm for giving me the opportunity to work on this research and for encouraging and guiding me during my work.

Foremost, I would like to thank Andreas Beck for sharing his broad and profound knowledge on valve trains and simulation and for improving this work with multiple discussions and contributing works.

Furthermore, I would like to thank all my former co-workers involved in the planning and execution of the tests and measurements in the context of this research for their high personal engagement and support.

Last but not least, I would like to thank all the people that supported me during this period regardless of the professional context. Especially I would like to thank my parents Katharina and René Pausin for encouraging me to pursue every goal I set myself. Furthermore, I would like to thank my friends for their endless understanding, loyalty and love during the past years.

October 2023 Angelina Eret

Acknowledgements

Abstract

The demand for more efficient combustion engines with fuel flexibility goes along with increasing component requirements regarding strength and durability, while the development costs should remain low. This applies in particular to combustion chamber adjacent components, which endure high temperatures and pressure loads. In this context, the present thesis focuses on the gas exchange valves of heavy-duty engines. Especially the valves on the exhaust side have an increased risk of fatigue failure due to their direct exposition to the hot exhaust stream.

The aim of this thesis is the generation of a predictive fatigue strength assessment to strengthen the frontloading of the exhaust valve design process and to increase the reliability of the component. The fatigue strength assessment requires the knowledge of all relevant loads of the component.

In this context, this thesis examines the loads of the exhaust valve during its working cycle. Beside the high temperature and cylinder pressure, further loads act on the exhaust valve. The actuation of the valve applies loads during valve opening and valve closing. The valve loads during closing reach higher levels due to an eccentric impact of the valve on the valve seat ring. Furthermore, a bold valve secondary dynamic in the form of valve bending vibrations is observed on the exhaust valves of heavy-duty engines increasing the valve load even more. The cause of this secondary dynamic is unknown.

This thesis investigates the valve loads to get the necessary input for the fatigue strength assessment. With respect to a predictive approach, the determination of valve dynamic and valve loads is based on a multibody simulation model of the valve train. In order to deliver predictive results and a transferable method, this simulation model includes all relevant physical effects to describe the valve dynamic accurately during all valve load phases of the working cycle.

The simulation model includes the flexible valve model and considers 3D-contacts of the valve with the valve guidance and the valve seat ring.

With the simulation model, the root cause for the bold valve secondary dynamic is examined iteratively. The model delivers not only the cause for the valve secondary dynamic but most importantly the critical valve loads. These loads deliver the input for the fatigue strength assessment.

To ensure the robustness of the load data determined by the simulation model the sensitivity of influences on the valve load is examined. In this context geometrical misalignment, fluctuations in load data and variable engine operation points are considered. A load collective based on the variation of influences on the valve load is the result of this analysis. All the results of the influence and sensitivity study are generated with the newly developed simulation model of the valve train.

The development of a measurement setup including the execution and evaluation of these measurements is another part of this thesis. In scope of these measurements are temperature and strain measurements of the valve performed on an engine testbed. The measurement results fulfil two purposes: Firstly, the data validate the simulation model of the valve train. Secondly, the statistical evaluation of the data delivers information on the distribution of valve loads. These findings are used in the subsequent fatigue strength assessment to increase the reliability of the results.

This thesis adds findings on the simulation of the valve dynamic to the current technological standards. Above all, it delivers a simulation-based and predictive approach on the fatigue strength assessment of four-stroke heavy-duty engine valves.

Zusammenfassung

Der Bedarf an immer effizienteren Verbrennungsmotoren, die flexibel in der Wahl des Kraftstoffes sind, führt zu stetig steigenden Anforderungen an die Festigkeit und Lebensdauer der Motorkomponenten. Dabei sollen die Kosten der Entwicklung möglichst gering bleiben. Insbesondere treffen diese Anforderungen auf brennraumangrenzende Bauteile zu, da diese besonders hohen Temperaturen und hoher Last durch den Zünddruck ausgesetzt sind.

In diesem Zusammenhang stellt die vorliegende Arbeit die Ladungswechselventile von Großmotoren in den Fokus, denn besonders die Auslassventile haben ein erhöhtes Risiko des Bauteilversagens aufgrund ihrer Exposition gegenüber dem heißen Abgasstrom.

Das primäre Ziel dieser Arbeit ist die Entwicklung eines prädiktiven Festigkeitsnachweises, um die Festigkeit des Auslassventils im Entwicklungsprozess frühzeitig berücksichtigen zu können. Dies mindert zum einen die Entwicklungskosten, zum anderen erhöht der prädiktive Festigkeitsnachweis die Zuverlässigkeit der Komponente. Für den Festigkeitsnachweis ist die Kenntnis aller Bauteillasten notwendig. Neben dem Zünddruck und der hohen Temperatur, die die Materialfestigkeit mindert, wirken weitere Einflüsse auf das Auslassventil. Die Aktuierung der Ventile induziert Kräfte in das Ventil, insbesondere beim Öffnen und Schließen des Ventils. Die Belastung des Auslassventils ist zudem erhöht durch ein exzentrisches Auftreffen des Ventils auf dem Ventilsitzring. Darüber hinaus kann an Auslassventilen von Großmotoren eine ausgeprägte Ventilsekundärdynamik beobachtet werden. Diese Sekundärdynamik äußert sich in einer Biegeschwingung des Auslassventils, die die Belastung des Auslassventils zusätzlich erhöht. Die Ursache dieser Sekundärdynamik ist unbekannt. Die Ventilbelastungen werden im Rahmen dieser Arbeit untersucht, um die notwendigen Eingabedaten für den Festigkeitsnachweis zu bestimmen. Um den prädiktiven Ansatz zu verfolgen, werden diese Untersuchungen anhand eines Mehrkörpersimulationsmodells des Ventiltriebs durchgeführt. Um dabei eine Übertragbarkeit der Modellierung und der gesamten Methode zur Festigkeitsbewertung sicherzustellen, enthält das Simulationsmodell alle relevanten physikalischen Effekte, die die Ventilbelastung und -dynamik beeinflussen. Zudem wird der gesamte Arbeitszyklus des Ventils mit allen Lastphasen abgebildet. Dabei enthält das Modell flexible Körper und dreidimensionale Kontaktmodellierungen der Kontakte von Ventil zu Ventilführung und Ventil zu Ventilsitzring.

Anhand dieses Modells werden die Hypothesen zur Ursache der Ventilsekundärdynamik überprüft. Iterativ wird so die tatsächliche Ursache bestimmt. Anhand dieses Modells können außerdem die Ventillasten als Basis für den Festigkeitsnachweis bestimmt werden.

Um die Robustheit der simulierten Lastdaten sicherzustellen und eine Varianz der Last über verschiedene Motoren zu berücksichtigen, wird die Sensitivität der Ventillast bezüglich verschiedener Einflussfaktoren untersucht. Zu den untersuchten Einflüssen gehören dabei geometrische Abweichungen, variierende Motorlastpunkte und variierende Kräfte auf den Ventiltrieb. Das Ergebnis dieser Analyse ist ein Lastkollektiv des Ventils, welches die Varianz der Einflussfaktoren berücksichtigt.

Ein weiterer Teil dieser Arbeit ist die Entwicklung und Durchführung von Messungen am Ventiltrieb. Die Messungen umfassen sowohl Temperatur- als auch Dehnungsmessungen am Ventil und werden am befeuerten Motorprüfstand durchgeführt. Die Messungen dienen zwei Zwecken: Zum einen kann mit den Messdaten das Simulationsmodell validiert werden. Zum anderen kann durch eine statistische Auswertung der Ventillastdaten die Verteilung dieser bestimmt werden. Diese Information beeinflusst den nachfolgenden Festigkeitsnachweis.

Die vorliegende Arbeit erweitert bekannte Methoden zur Simulation der Ventiltriebdynamik. Darüber hinaus wird ein prädiktiver, simulationsbasierter Ansatz zur Bewertung der Festigkeit von Viertakt-Großmotorventilen entwickelt.

Contents

Abbreviations

BDF	Backwards differentiation formula
CA	Crank angle
CCD	Central composite design
CFD	Computational fluid dynamics
CLT	Central limit theorem
CMS	Component mode synthesis
CVT	Complete valve train
DAE	Differential algebraic equation
DASSL	Differential algebraic system solver
DoE	Design of experiments
DOF	Degree of freedom
eFAST	Extended Fourier amplitude sensitivity test
EHD	Elastohydrodynamic
FAST	Fourier amplitude sensitivity test
FE	Finite elements
FEM	Finite element method
FKM	"Forschungskuratorium Maschinenbau"
FRM	Frequency response modes
GPD	Generalized Pareto distribution
IRM	Inertia relief modes
LEFS	Linear-elastic flexible spring
LHD	Latin hypercube design
MBS	Multibody simulation
MOM	Method of moments
MSM	Multi-segment measurement
OAT	One factor at a time
ODE	Ordinary differential equation
OVM	One valve model

PCM	Polygonal contact model
RBD	Random balanced design
SFS	Simbeam flexible spring
SSM	Single-segment measurement

Symbols

A	Virtual work of applied forces
$a_{inertial}$	Acceleration resulting from inertia relief
a_{BK}	Cyclic degree of utilization
A_c	Contact area
a_{damp}	Damping scaling factor
a_M	Constant of steel for mean stress sensitivity determination
$\boldsymbol{A}$	Rotational Matrix of the reference frame
$\boldsymbol{A}_{Sys}$	System matrix of the multibody system
B	Virtual work of mass accelerations
b	Elastic layer thickness
b_0	External surface forces
$\boldsymbol{B}_{Inp}$	Input matrix of the multibody system
b_M	Constant of steel for mean stress sensitivity determination
$\boldsymbol{C}$	Stiffness tensor
c_0	Constant of polynomial
C_{GT}	Contact stiffness of Greenwood-Tripp model
c_i	Coefficients of main effects
c_{ij}	Coefficients of interaction effects
c_l	Layer stiffness
$\boldsymbol{C}_{Sys}$	Output matrix of the multibody system
D_0	Strain tensor
d_0	External volume forces
$\boldsymbol{D}_{eq,modal}$	Damping matrix of structural viscous modal damping of the superelement
$\boldsymbol{D}_{\boldsymbol{FEM}}$	Damping matrix of the finite element model
$\boldsymbol{D}_{flex}$	Damping matrix of the flexible body
d_i	Modal damping values
d_l	Damping factor of the PCM contact
$\boldsymbol{D}_{modal}$	Damping matrix of modal damping of the flexible body
$\boldsymbol{D}_{\boldsymbol{SE}}$	Damping matrix of the superelement

$\boldsymbol{D}_{struc}$	Damping Matrix of structural damping
$\boldsymbol{D}_{struc,modal}$	Damping Matrix of structural modal damping
$\boldsymbol{D}_{inp}$	Feedthrough matrix of the multibody system
$\boldsymbol{D}_{v,modal}$	Damping matrix of viscous modal damping of the superelement
E	Young's Modulus
e^i	Unity vectors of reference frame
e^I	Unity vectors of initial frame
e_i	Unity vector
E_j	Elementary effects
e_j	Vector of the canonical base
E	Contact surface E
$\boldsymbol{F}$	System function
$F_{inertial}$	Force balancing the inertial load
F	Function of applied forces in a flexible body system
F_a	Function of applied external forces
f_a	Vector of applied forces
F_b	Boundary forces
$F_{c,N}$	Contact normal force
F_E	Contact force of the master surface in PCM contact
f_e	Force vector assembling the contributing components of every interconnection between the parts of the multibody system
F_{FEM}	Force vector of the finite element model
F_i	Interior forces
f_i	Frequency of the mode i
F_N	Normal force of the PCM contact
F_{pcm}	Contact force of an element of the master surface in PCM contact
F_{SE}	Force vector of the superelement
F	Contact surface F
F_T	Tangential force of the PCM contact
F_u	Mean excess function
$f_{W,\sigma}$	Fatigue strength factor for completely reversed normal stress
$f_{W,\tau}$	Fatigue strength factor for completely reversed shear stress
$\boldsymbol{g}$	Function describing the constraints of the multibody system
G_ξ	Generalized Pareto distribution
h	Lubrication gap function
h_n	Step size
h_*	Nominal lubrication film thickness
$\boldsymbol{I}$	Identity matrix
j_F	Material safety factor
$j_{n,S}$	Statistical conversion factor
$\boldsymbol{J_R}$	Jacobian matrix of virtual rotational displacements
j_D	Fatigue safety factor
j_S	Load safety factor
$\boldsymbol{J_T}$	Jacobian matrix of virtual translational displacements
K	Stiffness matrix of the finite element body to be reduced

k	Slope of the S-N curve
K_{AK}	Mean stress influence factor
K_{BK}	Amplitude strength factor
K	Elastic modulus
$\boldsymbol{K_{FEM}}$	Stiffness matrix of the finite element model
k_{mat}	Slope of S-N material curve
$\boldsymbol{K_{SE}}$	Stiffness matrix of the superelement
$K_{T,D}$	Temperature factor
$K_{WK,\sigma}$	Design factor of component under normal load
$K_{WK,\tau}$	Design factor of component under shear load
L	Angular momentum
l_i^k	Basis polynomials
LSD	Logarithmic standard deviation
$\boldsymbol{M}$	Mass matrix
x	Positional vector including translational and rotational degrees of freedom
M_E	Contact torque of the master surface in PCM contact
M_{ee}	Mass matrix with elastic contribution
M_{et}	Mass matrix with elastic-translational contribution
m_F	Mean of material data distribution
M_{FAST}	Order of interference free frequencies in FAST
$\boldsymbol{M_{FEM}}$	Mass matrix of the finite element model
M_{pcm}	Contact torque of an element of the master surface in PCM contact
M_{rr}	Mass matrix with rotational contribution
M_{rt}	Mass matrix with rotational-translational contribution
$\boldsymbol{M_{SE}}$	Mass matrix of the superelement
M_σ	Mean stress sensitivity on normal stress
M_τ	Mean stress sensitivity on shear stress
M_{tt}	Mass matrix with translational contribution
N	Cycle number
$\bar{N}$	Assigned cycle number
N_D	Knee point cycle number
$\boldsymbol{N}$	Force distribution matrix
n_f	Number of factors
n_l	Number of factor levels
n_M	Number of random variables
n_{pcm}	Normal vector of the master surfaces in PCM contact
N_{tot}	Total number of samples
N_u	Number of samples over a threshold
O^i	Origin of reference frame
O^I	Origin of initial frame
P	Probability
μ_{vis}	Lubrication viscosity
p	Linear momentum
$P_{A,log}$	Failure probability for logarithmic normal distribution

$P_{A,N}$	Failure probability for normal distribution
p_{GT}	Contact pressure of Greenwood-Tripp model
p_N	Nominal position satisfying positional constraints
P_{oil}	Oil film hydrodynamic pressure
p_{poly}	Interpolation polynomial
P_S	Survival probability
p_{SE}	Harmonically time-dependent unit load
q	Vector of generalized forces and torques
R_0	Position of a point in the body reference configuration
R^2_{det}	Coefficient of determination
R_{jour}	Journal radius
R_m	Material's tensile strength
R_σ	Load ratio
S	Vector of infitesimal rotation
$\mathbf{S}$	Rotational tensor
S_F	Fatigue strength safety factor
s_F	Standard deviation of material data distribution
s_f	Control variable of FAST
$s_{flx,i}$	Modal coordinates
σ_m	Mean stress
σ_{max}	Maximum stress
σ_{min}	Minimum stress
σ_{WK}	Component specific fatigue limit for normal load
s_σ	Transient component stress
$S_{X,50\%}$	Geometric mean value of material stress data
$S_{X,97.5\%}$	Material stress data for survival probability $P_S = 97.5\%$
T	Torque
t	Time
τ_{WK}	Component specific fatigue limit for shear load
$\boldsymbol{T}_{CMS}$	Transformation matrix of component mode synthesis
$\boldsymbol{T}_{Damp}$	Transformation matrix of the modal structural damping matrix
$\boldsymbol{T}$	Transformation matrix
u	Deformation vector of the flexible body
u_0	Relation of mean and standard deviation
u_{Inp}	Input vector of the multibody system
u_N	Displacement in contact normal direction
u_{tr}	Threshold of a distribution
v	Translational velocity
v_ϵ	Minimum velocity for Coloumb friction model
v_N	Normal component of contact velocity
v_T	Tangential component of contact velocity
w	Vector of translational degrees of freedom
W_s	Virtual work of strain
X	Random variable
x_b	Displacements on interface nodes (boundary displacements)

X_f	Vector of inputs
x_f	Variables of the polynomial
x_i	Displacements on interior nodes
X_n	Random variable
$\bar{X}$	Mean value of a random variable
x_{State}	State vector of the system
Y	Output attribute
y	Vector of generalized coordinates
$\bar{Y}_i$	Mean of output values
$\hat{Y}_i$	Output values of interpolation polynomial
y_{out}	System output
z	Vector of generalized velocities
z_I	Positional state vector
z_{II}	Positional velocity vector
Z_n	Standardised random variable
α^C_{k-i}	Coefficient of corrector equation
α_i	Critical damping factor of the mode i
α^P_{k-i}	Coefficient of predictor equation
β_i	Stiffness proportional damping factor of the mode i
β^C_k	Coefficient of corrector equation
Γ	Vector function describing holonomic constraints
δ_{opp}	Surface roughness of opposite side
δ_{ref}	Surface roughness of reference side
$\delta_{r,o}$	Combined surface roughness of reference and opposite surface
Δ_s	Increment size of Morris sample
$\boldsymbol{\epsilon}$	Strain tensor
ϵ	Model error
ξ	Form parameter of the generalized Pareto distribution
η	Second Lamé constant
τ	Increment of Jacobian iteration
Λ	Stress tensor
λ	Lagrange multipliers
μ	Friction coefficient
μ_M	Expected value of the normal distribution
$\tilde{\mu}$	Ratio of samples over threshold and total samples
ν	Poisson's ratio
Ω	Harmonic response frequency
ω	Rotational velocity
$\omega_{f,i}$	Factor frequencies of FAST
ω_j	Eigenfrequency
$\tilde{\omega}$	Angular velocity with respect to reference frame
$\boldsymbol{\Phi}$	Mode shape matrix
$\boldsymbol{\phi_C}$	Constraint mode matrix
ϕ_i	Eigenvectors of the superelement
ϕ_j	Eigenvectors of the finite element model

$\boldsymbol{\phi}_N$	Constrained normal mode matrix
$\prod_k$	Polynomial interpolating k solution points
$\boldsymbol{\psi}$	Vector function describing non holonomic constraints
ρ_0	Material density
$\boldsymbol{\sigma}$	Stress tensor
$\sigma_{\mathrm{a},1}$	Maximum stress amplitude of load spectrum
σ_δ	Standard deviation of surface roughness distribution
σ_{equi}	Equivalent stress amplitude
σ_M	Standard deviation
$\sigma_{r,o}$	Standard deviation of reference/opposed surface roughness distribution
$\tilde{\sigma}$	Estimated standard deviation of the generalized Pareto distribution
$\sigma_{W,zd}$	Stress limit of completely reversed normal stress
$\tau_{W,s}$	Stress limit of completely reversed shear stress
θ	Switching function of discontinuity handling
ζ_i	Eigenvalues

List of Figures

List of Tables

Chapter 1
Introduction

The history of large bore internal combustion engines started with Rudolf Diesel's patent from 1892. The first diesel engine was running five years later. Nowadays, medium and low speed engines are the predominant propulsion system of ships. In combination with an electric generator they provide electric energy in power plants and as gensets in ships.

The immense impact of climate change increases the demand for engines with increased efficiency and also for engines that run on alternative fuels and therefore minimize carbon dioxide emissions.

These requirements lead, among other things, to increasing cylinder pressures, increased engine charge air pressures and increased temperatures. Alternative fuels can also lead to a reduction of strength in materials.

It is therefore mandatory to have a profound knowledge of the material behaviour and the mechanical loads of all highly stressed engine components so that a reliable fatigue strength assessment for each component can be carried out. With the knowledge of physical influences on each component's strength, changes in loads and materials can be considered efficiently.

This thesis focuses on the gas exchange valves of medium speed four-stroke engines. Usually, those engines have two inlet and two exhaust valves per cylinder. The inlet valves are supposed to open during intake stroke to enable an influx of fresh air. For all other strokes the valve is supposed to seal the combustion chamber. The exhaust valves, on the other hand, open during the exhaust stroke to enable the exhaust gas to leave the combustion chamber. These valves are supposed to seal the combustion chamber and remain closed during the other strokes. The valves are a mandatory component. Hence, a failure of the valves disables the operation of the engine.

Due to their functionality and position in the combustion chamber, the valves are exposed to several mechanical loads. In addition, the high temperatures decrease

A. Eret, *Dynamic and Fatigue Assessment of Heavy-Duty Engine Valves*, Mechanics and Adaptronics, https://doi.org/10.1007/978-3-031-49151-1_1

the valve material strength. This combination of effects on the valve has a negative influence on the valve fatigue strength.

The high actuation velocity of the valves leads to significant dynamic effects. To determine the mechanical valve loads resulting from these effects, it is mandatory to determine the causes and manifestations of the valve dynamic. The mechanisms leading to dynamic loads in the valve are outlined in further sections of this chapter.

This thesis presents the simulative approach to determine the dynamic behaviour and dynamic loads of the gas exchange valves. These loads then are input of a fatigue strength assessment of the valve. This fatigue strength assessment relates the present loads of the valve to the valve strength and therefore assesses the probability of failure of the valve design during its operation period.

1.1 Main Thesis Aim

This section gives a brief overview over the main thesis aims and the general approach to achieve these aims.

The present thesis aims for the fatigue assessment of the valves of heavy-duty four-stroke engines and in this context for the determination of the valve loads.

The valve design assessment at MAN Energy Solutions (MAN ES) is based on measurement data of the valves. This approach relates the component strength to the component load as previously described. In contrast, an approach based on data predictively determined by a virtual twin of the component delivers loads based on the considered physical effects.

For the measurement based method the strength value is determined by measured loads of failed valves with a certain safety margin. Measurements of valve strain during engine operation deliver the valve load. But the amount of strain gauges on the valves is limited. Thus, strain gauges are only applied to the most critical area, which is assumed to be the radius between valve stem and valve plate. The strain measurements in this area show high strain values close to the empirical limit of valve strain.

This measurement based method clearly shows deficits: The limit value is an estimation only roughly considering physical influences. Changes in temperature, component geometries, pressures etc. are not considered directly and cause a high uncertainty during the design process. The empirical approach is furthermore resource-consuming because changes in the valve design can not be assessed before the valve is fully manufactured and applied to the engine, even if the changes are small. Thus, there is a lack in frontloading during the valve design process.

The predictive and simulation-based fatigue assessment is able to reduce uncertainties in the design process and increase the frontloading to reduce time and cost of the valve design process. But this method requires further steps to determine the valve loads.

The valve loads are influenced by the high valve dynamic. The dependency of the valve dynamic from boundary conditions of engine operation is unknown. From this, the additional aim of an analysis of the valve dynamic can be derived. The analysis includes the causes of the valve dynamic, the determination of influences on this valve dynamic and the impact of the valve dynamic on the valve load. This analysis builds the base for the general understanding of the complex valve load.

To reduce the development and measurement costs and to enable an early reliability check of the component in the design process a further aim of this thesis is the generation of digital model that includes the relevant physical effects of the valve or even a digital twin.

So far, simulation models of the MAN ES valve train focus on the validation of cam contours not the valve dynamic. Therefore, the modelling depth regarding the valves is low and includes several simplifications. With the focus of the model shifting to the valve, less important components like the cam, the pushrod and the rocker arm are left out. Instead, the simulation approaches for the valve and its direct contact partners are more detailed and able to deliver reliable valve load results. This model is called OVM and shown in more detail in chapter four.

Focusing on the fatigue assessment the distribution of valve loads is a further important aspect. The fatigue assessment requires safe load levels. In this context, the term safe relates to the presumption that an exceeding of the load level occurs very rarely during the component's lifetime. The absolute number of tolerated exceeding of the fatigue limit depends on the tolerated probability of the limit exceeding. Usually, this is ensured by analysing the load distribution and determining a high quantile load level from the distribution parameters. Therefore, the analysis of the valve load distribution is an obligatory part of the simulation based fatigue assessment.

In the following, the next subchapters give a more detailed overview over the valve train types this thesis focuses on. Moreover, the valve loads and valve dynamic are described. Finally, the chapter gives an overview over the structure of the present thesis.

1.2 Nomenclature

The present section briefly explains the nomenclature of engines and introduces the valve train types examined in this thesis.

The naming of the considered MAN ES engines provides information on their geometry. The first number in the name corresponds to the bore diameter in centimetres, while the second number describes the piston stroke, also in centimetres. The number of cylinders is also part of the engine name. The configuration of the engine is described by a letter following the number of cylinders. A 'L' describes a configuration with all cylinders of the engine being in line. The 'V' configuration consists of two banks of cylinders. The banks are arranged in a way that they form the letter 'V'. Other letters may be added to the name that describe the fuels the engine is operating on. 'D' stands for Diesel, 'G' stands for gas and 'DF' stands for dual fuel, which means that the engine can operate either on diesel or gas. The engine shown in Fig. 1.1 is a 12 V 51/60 DF. It therefore has two banks with 6 cylinders each

Fig. 1.1 12V 51/60 DF engine

in a V-configuration. Each cylinder has a diameter of 510 mm (51 cm) and a piston stroke of 600 mm (60 cm). The engine runs either on gas or diesel and is therefore a dual fuel engine (DF).

Usually, the valve train construction is the same for all engine configurations with the same bore diameter and stroke. Thus, for the sake of simplicity, engines are mainly referred to by these two numbers, e.g. 51/60. All configurations are included by this reference.

1.3 Load Exchange Valves of Heavy-Duty Engines

This section introduces the valve train types of heavy-duty four-stroke engines. The actuation principle of the valve determines the dynamic parameters and also the number of overall valve train components that possibly influence the valve dynamic. The considered valve train types are outlined in the first part of this section. The second part of this section focuses on the valve loads that occur for the prior specified valve train types.

The inlet and exhaust valves are part of the valve train of internal combustion engines. The shape of the valve lift curve depends mainly on the contour of the cam but is also influenced by the valve train kinematics, elasticities and dynamic behaviour of the valve train or specific valve train components. The most commonly known application occurs in combustion engines of passenger cars. Regarding mechanical valve actuation principles almost all of those engines use overhead camshafts (OHC). A subtype of this construction is the double overhead camshaft (DOHC) [1, 2].

However, top-mounted camshafts are not the common solution for heavy-duty engine valve actuation. Because of the big distances between crankshaft and camshaft, the bottom-mounted camshaft is the common material saving construction of heavy-duty valve trains. This valve train type goes along with a more complex dynamic behaviour because it has more components and therefore more complex contacts than the valve trains with top-mounted camshafts.

1.3.1 Structure and Functionality

Two different designs of valve trains differing in the rocker arm design are considered in the thesis. In one case, the rocker arm acts directly on the valves. In the other case, the rocker arm acts on the valve bridge. In both designs the cam is in contact with the follower that in turn drives the pushrod due to the actuation of the cam. The pushrod actuates the rocker arm.

For the valve train design shown in Fig. 1.2, the rocker arm transmits the pushrod actuation due to its tilting movement to the valve bridge. A small thrust piece on the rocker arm side, called elephant foot, translates the rotating movement to a translational movement. A spherical joint on the rocker arm side and a plane surface on the valve bridge side allow a sliding movement of the elephant foot on the valve bridge. Between elephant foot and valve bridge a clearance is adjusted to avoid prestressed operation. The valve bridge transmits the vertical movement from the rocker arm to two valves simultaneously. This design is used for many MAN ES engine types. This thesis focuses on the valve train of the 49/60 engine which is the same type as the 45/60 engine valve train.

The main difference between the above described valve train design and the valve train design, shown in Fig. 1.3, is the rocker arm geometry. In this design, the rocker arm acts directly on the valve, leaving out the valve bridge. Two elephant feet are in contact with the valves, sliding on their upper surfaces. Valve clearance is adjusted between the valves and the elephant feet.

1.3.2 Valve Loads

The exposition of valve train components, the exhaust valve in particular, to temperatures up and above 500 °C decreases the material strength significantly. The temperature level of inlet valves is significantly lower. Therefore, temperature related failure mechanisms occur on exhaust valves more often.

The valves are dynamically loaded by the excitation of the cam that is transmitted through the oscillating valve train system. Moreover, they are exposed to cylinder pressure. The valve has three main load phases during an engine working cycle.

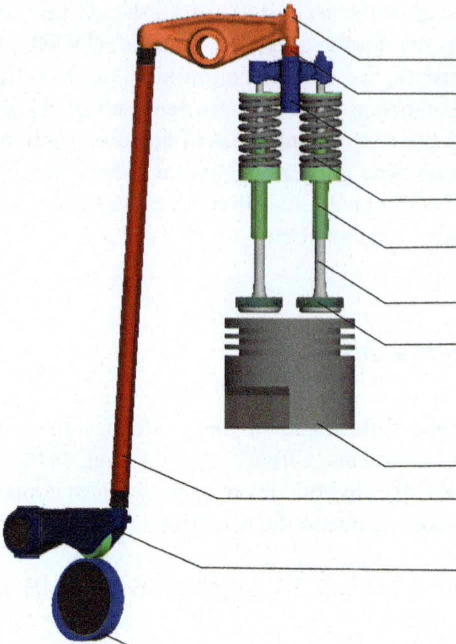

Fig. 1.2 Valve train type 49/60 engine and 45/60 engine

The first phase is the ignition phase. During ignition, all valves of the cylinder are closed and pressed into their valve seat rings due to the increasing pressure in the combustion chamber. Therefore, normal loads are induced in the valve.

The second phase is the valve opening. During this phase the valve is actuated by the rocker arm or valve bridge. The necessary force to open the valve induces normal loads to the valve. The actuating force needs to overcome the cylinder pressure, which amounts to up to 20 bar at valve opening, and the valve spring force. The higher the actuating force, the higher the induced normal loads in the valve.

The third phase is the valve closing. While closing, the valve spring pulls the valve back to its valve seat ring and therefore transmits normal force to the valve. The valve load resulting from the impact of the valve on the valve seat ring increases with valve speed. Therefore, the valve speed right before closing is limited to a certain level in the design process.

In addition, the first contact of valve and valve seat ring is pointed leading to bending loads in the valve. The pointed contact is the result of a slight misalignment between valve and valve seat ring. The higher the misalignment between valve seat ring and valve, the higher the induced bending load in the valve. Figure 1.4 shows the mechanical loads of the valve.

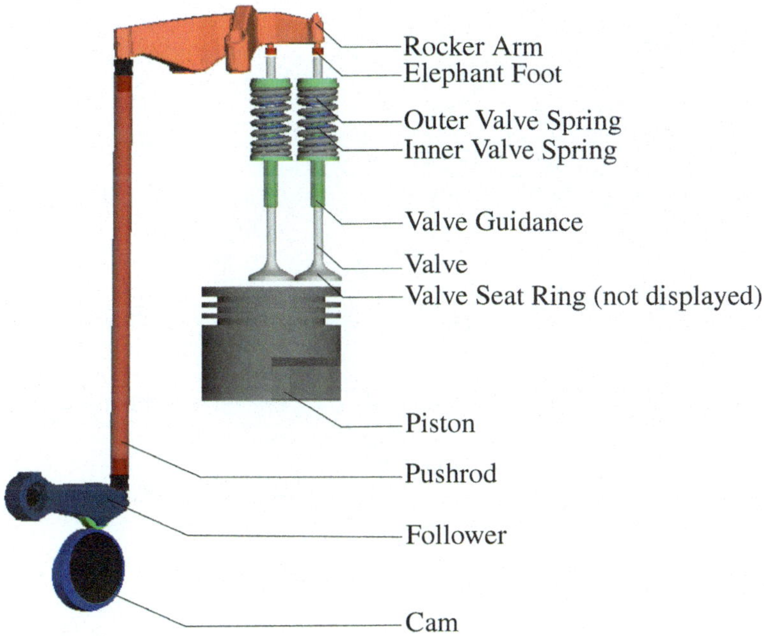

Fig. 1.3 Valve train type 48/60 engine

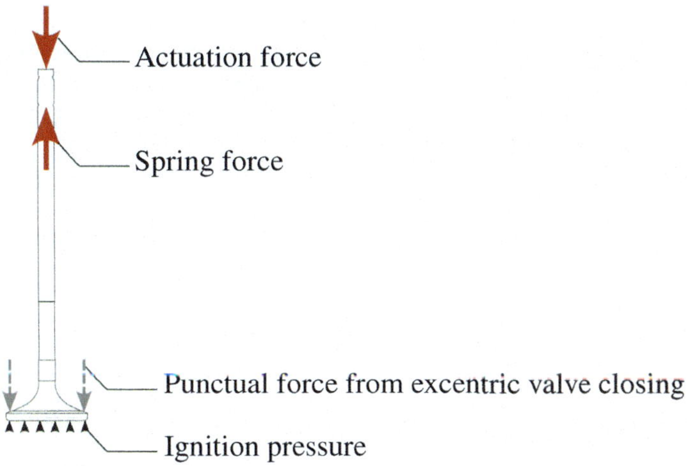

Fig. 1.4 Valve loads during working cycle

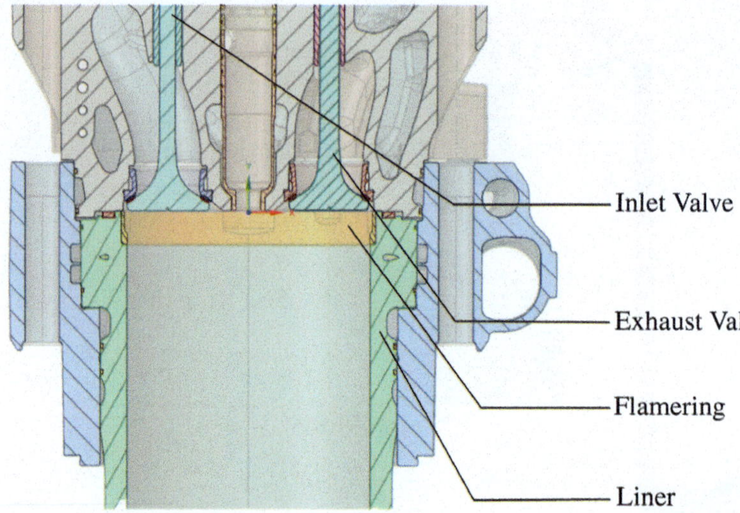

Fig. 1.5 Cross-section of cylinder head and liner

A known mechanism that may lead to an increased misalignment of the valve is the valve secondary dynamic. The term secondary dynamic refers to a movement of the valve perpendicular to its main direction of movement. The secondary dynamic results in a transverse bending during the valve opening phase. The fact that exhaust valves undergo significant secondary movement can be seen by two observations made at several engines. One of these observations is contact marks on the valve flame ring in various engine types. The position of the flame ring in the liner is shown in Fig. 1.5. In the pictures in Fig. 1.6 showing the contact marks on the flame ring it becomes clear that the contact of valve and flame ring only occurs on the exhaust side. The valve plate is at the same height in the cylinder as the flame ring when it is open. From the position of the marks on the flame ring it can be deduced that contact must have occurred when the valves were fully open. Consequently, the valves must have overcome the transverse distance to the flame ring of 3 to 5 mm, depending on the engine type. From that, a bending vibration of the valve can be assumed.

Further evidence of the valve bending can be observed in strain measurements made in the valve radius and in the middle of the valve stem. Figure 1.7 shows strain signals from two strain sensors that are positioned in the same horizontal plane but have an angular distance of 180° to each other. The signals show a clear bending characteristic with opposing peaks. Valve bending also causes increased wear due to increased mixed friction in the lower part of the valve guidance. In large-sized engines a preferred direction of guidance wear can be observed which leads to an ovalised geometry in the lower guidance part.

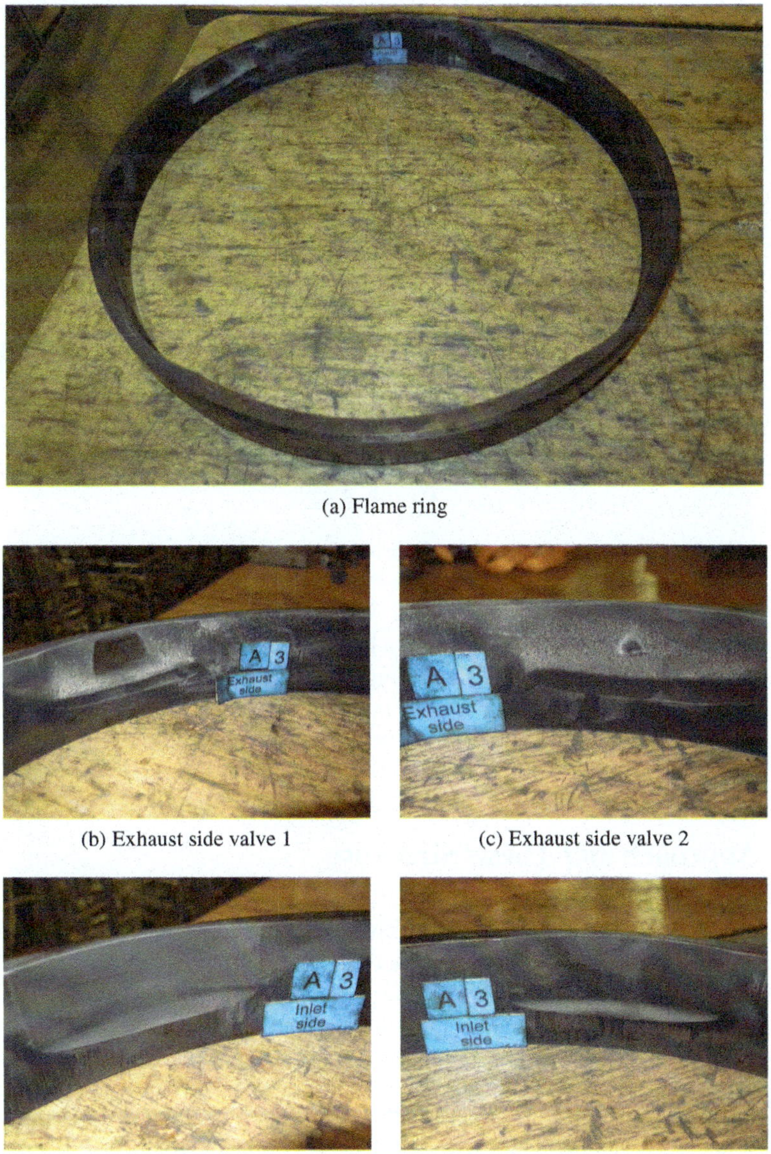

(a) Flame ring

(b) Exhaust side valve 1

(c) Exhaust side valve 2

(d) Inlet side valve 3

(e) Inlet side valve 4

Fig. 1.6 Contact marks on flame ring

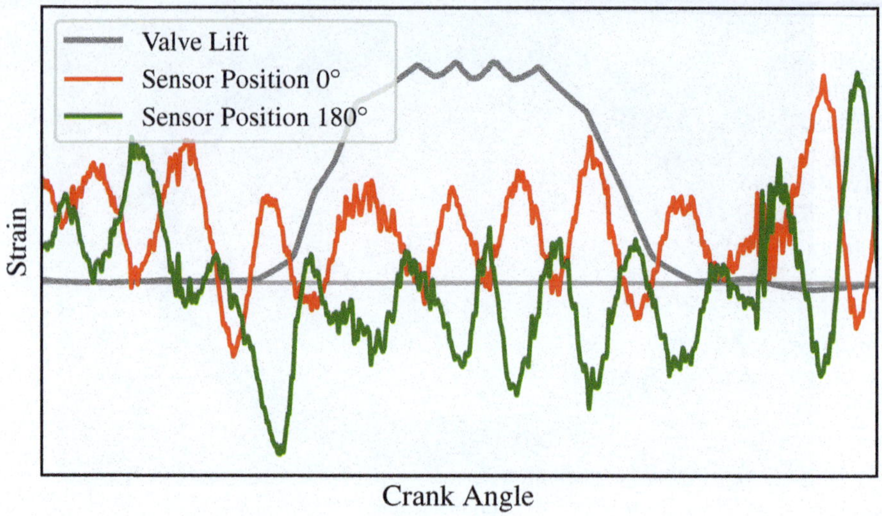

Fig. 1.7 Bending signals from valve strain measurements

In summary, the exhaust valves show a higher secondary dynamic observed from the strain measurements and the flame ring contact and additionally are exposed to much higher temperatures compared to inlet valves. Therefore, this thesis focuses on the exhaust valves as they are more critical regarding fatigue strength.

1.4 Approach and Thesis Structure

This section specifies the approach of the investigations on valve dynamic and valve fatigue strength assessment. In this context, the parameters most likely to influence the valve dynamic during engine operation are pointed out. The second part of this section gives an overview over the structure and further topics of the present thesis.

Section 1.3.2 points out the fact that gas exchange valves and exhaust valves in particular are highly stressed components. A lot of factors during engine operation increase the valve load even more. These include geometric misalignments and displacements. Some of these misalignments or displacements occur due to manufacturing tolerances. Others occur due to thermal deformations of the cylinder head during engine operation. Both kinds of deformation can result in an eccentricity of the valve's and valve seat ring's rotational axes increasing valve bending load during valve closing.

Furthermore, the valve experiences both dynamic and static flow forces in several directions. The exhaust valve opens when the cylinder pressure is still at up to 20 bar and thus enables a directional flow passing by the exhaust valves.

External forces, geometrical tolerances and misalignment and further known physical effects influencing the valve dynamic are included in the simulation model. The investigation of influence factors is linked to the determination of the cause for the valve bending vibration that was observed in strain measurements and flame ring damage report. With the detailed knowledge on the causes and characteristics of the valve dynamic it is possible to deduce the actual valve loads during engine operation and to use these for a subsequent fatigue strength assessment.

In general, there are two methods of valve dynamic examination. One option is the experimental examination of the valve dynamic. This examination requires a huge number of test runs to find all the relevant influence factors and analyse their root causes. This experimental approach leads to particularly high costs because of a heavy-duty engines high fuel consumption, high cost due to the complex sensor application and the general test bed costs. Furthermore, the fully engaged engine testbeds would be unavailable for any other measurements. From an economic perspective, this experimental approach is unreasonable.

The second option for an approach of valve dynamic examination is the simulation based method that was introduced in the beginning of this chapter. Therefore, the simulation model of the valve and the relevant neighbouring components need to include the relevant influencing factors and display the real valve train behaviour as accurate as possible. There are two simulation methods suitable to investigate component dynamic and loads, the multibody simulation (MBS) method and the finite element method (FEM). Both methods can be found in publications on valve train dynamic investigations. These publications are examined in the third chapter of this thesis.

The method chosen for the investigations carried out in this thesis is the MBS method. The main reason is the option to simulate not only single load events during the valve working cycle like the valve opening or valve closing but to simulate the whole working cycle or even a series of working cycles. Moreover, the MBS method allows a broad spectrum of contact modellings and opening and closing of contacts between two components. This is assumed to be a more accurate representation of the real valve train behaviour. In addition, the method allows the implementation of flexible bodies, which is mandatory for the component load determination.

For the purpose of simulation model validation measurement data from test bed experiments are needed. However, the costs for these measurements are far lower than for a fully experimental approach. The measurement set up delivers data to a statistical consideration of valve loads influencing the following fatigue strength assessment. The set up and execution of these measurements is a further part of this thesis.

The simulative approach delivers a validated model that enables an early assessment of dynamic and fatigue limit of the valve. Overall, the simulative approach is a resource saving process that enables a rapid adaption of variable valve train configurations and changes in relevant boundary conditions.

The thesis starts with an overview over the theoretical principles and general methods of MBS, statistic evaluation methods and fatigue strength assessment. The following chapter outlines the current state of the art regarding valve dynamic simulation, measurement approaches on valve dynamic analysis and valve fatigue strength assessment. The fourth chapter of this thesis describes the structure of the MBS model of the valve. The influences on the valve train potentially affecting the valve dynamic are outlined. Furthermore, the chapter presents hypotheses on the cause of the valve secondary dynamic and the iterative process of determination of the valve secondary dynamic root cause. Finally, the setup, the processing and the evaluation of the valve train measurements are explained. Finally, the chapter shows the measurement based validation of the MBS model. The fifth chapter shows the application of the findings of the previous chapter. The validated simulation model builds the base for investigations carried out in this chapter. Before assessing the fatigue limit, the sensitivity of the valve load towards potentially influencing factors is investigated. With the application of design of experiments (DoE) on the simulation model statistical quantities of the valve load distribution are estimated and taken as a base for the subsequent fatigue strength assessment that is also outlined in the fifth chapter.

The final chapter gives a conclusion of the findings of this thesis and provides an outlook on further steps in the investigation of valve dynamic and strength.

References

1. MAHLE GMBH: Aufgabe von Ventiltriebsystemen. In: *Ventiltrieb*. Wiesbaden : Springer Fachmedien Wiesbaden, 2013 , 1–6
2. XIN, Q. : *Diesel Engine System Design*. Cambridge, England : Woodhead Publishing, 2011

Chapter 2
Theoretical Principles

The following chapter delivers the preliminaries to the method developed within this thesis by introducing the theoretical background of the techniques used in the fields of MBS and statistics. Furthermore, the third section of this chapter describes the assessment of the valve fatigue strength following [34]. As mentioned in the introduction chapter, the MBS is the chosen simulation method to determine the valve dynamics and deliver the load input for the fatigue assessment. The statistical analysis of the valve load data whether they are from simulation or measurement ensures that the load level chosen for fatigue assessment is safe and excesses of this load level have a sufficiently small probability. The statistical analysis of valve load data therefore enables the detection of sensitivities of the valve load towards influencing parameters. The fatigue assessment delivers the desired evaluation of the valve design under the given operational boundary conditions.

The section describing the method of MBS starts with the equations of motion that describe either constrained or unconstrained mechanical systems consisting of multiple rigid bodies including interactions or external forces to explain the basic mechanical principles of the method. Moreover the dynamic of flexible bodies in MBS and the implementation of such bodies in MBS models is outlined. Flexible bodies are necessary to determine component loads. Furthermore, the section describes advanced contact modelling approaches. The contact models are important to describe the interaction between two or more bodies in the resulting model.

The section addressing statistics gives a brief introduction to the widely used Gaussian distribution, before focusing on the extreme value distribution. The aim of the distribution determination is the description of the expectable valve load maximum. Moreover, the section briefly summarizes statistical methods of measurement data evaluation. Furthermore, methods of DoE are introduced. These methods enable the analysis of valve load sensitivities and the analysis of valve load variance.

Finally, the last section gives on overview over the fatigue strength assessment according to “Forschungskuratorium Maschinenbau” (FKM).

A. Eret, *Dynamic and Fatigue Assessment of Heavy-Duty Engine Valves*, Mechanics and Adaptronics, https://doi.org/10.1007/978-3-031-49151-1_2

2.1 Multibody Simulation

The MBS is a numerical simulation method for the description of dynamical system's behaviour. Such a dynamical system can consist of rigid or flexible bodies connected to each other either by forces, joints or constraints. The latter two decrease the number of degree of freedom (DOF) of the considered system.

Contact between the bodies of the system can be described by several formulations of force element ranging from simple spring-damper elements to complex elastohydrodynamic (EHD) contact formulations.

The fields of application for MBS are manifold and its importance is increasing due to the ability to simulate complex motion sequences of mechanical systems [23]. It is widely used in automotive engineering e.g. in the simulation of chassis dynamic. Another application is the field of machine tool control, where MBS is used to optimise new approaches in efficient machine design or active damping of machine vibrations [9]. Also the science of sport, biomechanics in particular, uses MBS to simulate the human body and its dynamic [16].

A summarising overview over the methodology of MBS for rigid and flexible bodies, its analytical and recursive formalisms and application examples is given by [38].

2.1.1 *Equations of Motion*

An MBS model simulates a mechanical system consisting of several bodies interacting with each other. The mathematical description of mechanical systems is provided by the equations of motion. This subsection shows the derivation of these equations.

There are several approaches on the mathematical description of MBS systems, introduced in e.g. [9, 36] that are referenced within this section. A common formulation of the equations of motion of MBS systems is Lagrange's equation of the first kind. This section will lead to this formulation of the equations of motion.

The mechanical laws defined by Newton and Euler describing the dynamic of a rigid body relating to its centre of gravity form the basis of the equations of motion describing a multibody system. Newton's law says that the rate of change of a body's linear momentum p is equal to the sum of all external forces F_a acting on the body (2.1.1). In addition, the principle of angular momentum, firstly presented by Euler (2.1.2), says that the rate of change of a body's angular momentum L^o, regarding the space point o is equal to the sum of all external torques T acting on that body [36].

$$\frac{d}{dt}p = F \tag{2.1.1}$$

$$\frac{d}{dt}L^o = T \tag{2.1.2}$$

The formulation of an unconstrained, planar system with the mass matrix $\boldsymbol{M}$ can directly be derived from the Newton-Euler equations with the positional vector x and its derivatives $\dot{x}$ and $\ddot{x}$, describing the body's velocity and acceleration. F_a is the function of all external forces. The variable t is the time. The resulting equation (2.1.3) is of type ordinary differential equation (ODE) [11].

$$M\ddot{x} = F_a(t, x, \dot{x}) \tag{2.1.3}$$

Starting from this, the present section will lead to the general formulation for three-dimensional, constrained multibody systems for rigid bodies. Therefore, the d'Alembert principle, describing a constrained mechanical system, will be introduced later on in this section.

Firstly, this section introduces the general elements of a MBS system. Every MBS system consists of an arbitrary but nevertheless finite number of bodies. For this section, each body is assumed to be a rigid body. Each body has six DOF for an unconstrained system consisting of three translational DOF, considered by the vector w in (2.1.4), and three rotational DOF. The 3D-rotation depending on α, β and γ is denoted in the rotational tensor S shown in (2.1.5). Euler angles or Kardan angles can be used to describe the 3D-rotation of a body. Another option is the usage of Euler parameters also known as quaternions. The quaternions avoid the disadvantage of singularities, also known as Gimbal Lock, for specific angles of Euler and Kardan angles. In addition, quaternions are unique [36].

$$w = [w_1, w_2, w_3]^T \tag{2.1.4}$$

$$\boldsymbol{S} = \boldsymbol{S}(\alpha, \beta, \gamma) \tag{2.1.5}$$

The positional vector x introduced in (2.1.3) sums up both the translational and the rotational coordinates and is therefore used in further explanations to describe the systems DOF.

The total number of DOF k for a mechanical system is decreased by the introduction of constraints e.g. kinematic relations between bodies of the mechanical system. There are several types of constraints or ways these constraints can be introduced to the system, e.g. joints, contact formalisms or boundary conditions.

In a mechanical system consisting of multiple bodies, the movement of bodies related by kinematic constraints is not independent from each other and the bodies' coordinates aren't either. Therefore, a body's positional coordinates can be denoted only by the coordinates describing the independent system movements instead of a full set of coordinates. Here, the position vector using these so-called generalised coordinates is denoted by y in (2.1.6). For this notation, all dependent coordinates are transformed into terms depending on the generalised coordinates that are the independent coordinates of the system. The minimal number of generalised coordinates corresponds to the number of overall system DOF [28].

$$y = [y_1, y_2, \ldots\ldots, y_k]^T \tag{2.1.6}$$

With y the application of holonomic, rheonomic constraints can be expressed explicitly by (2.1.7). This equation defines the generalised coordinates and thus fulfils the constraints. The constraints can also be denoted implicitly by (2.1.8) with the vector function Γ describing the equations between coordinates to fulfil the constraints. Therefore, the number of columns in Γ equals the number of constraints.

$$x = x(y, t) \tag{2.1.7}$$

$$\Gamma(x, t) = 0 \tag{2.1.8}$$

With respect to the inertial frame, the velocity vectors for translational velocity v and rotational velocity ω can now be derived from positional coordinates using the Jacobian matrices $\boldsymbol{J_T}$ characterizing the virtual translational and $\boldsymbol{J_R}$ characterizing the virtual rotational displacement of the system. $\tilde{v}$ and $\tilde{\omega}$ are the local velocities. The vector S is the vector of infinitesimal rotation derived from $\boldsymbol{S}$ [37].

$$v = \frac{\partial w}{\partial y}\dot{y} + \frac{\partial w}{\partial t} = \boldsymbol{J_T}(y, t) + \tilde{v}(y, t) \tag{2.1.9}$$

$$\omega = \frac{\partial S}{\partial y}\dot{y} + \frac{\partial S}{\partial t} = \boldsymbol{J_R}(y, t) + \tilde{\omega}(y, t) \tag{2.1.10}$$

The derivation of the translational and rotational accelerations follows the same scheme as the derivation of velocities.

For the reason of non-integrability of non-holonomic constraints they are considered in the velocities as they reduce the number of the system's motional DOF. With the vector z characterizing the systems' generalised velocities the g non-holonomic constraints, analogue to holonomic constraints, are denoted explicitly by (2.1.12) and implicitly by (2.1.13) with the vector function Ψ describing the relations between the coordinates y considering the non-holonomic constraints.

$$z = [z_1, z_2, \ldots\ldots, z_g]^T \tag{2.1.11}$$

$$\dot{y} = \dot{y}(y, z, t) \tag{2.1.12}$$

$$\Psi(y, \dot{y}, t) = 0 \tag{2.1.13}$$

The vector q including all generalised forces and torques is introduced [37].

$$q = [f_1^T, f_2^T, ..., l_1^T, ..., l_p^T]^T \tag{2.1.14}$$

Now, the d'Alembert principle is applied to denote the equations of motion for the whole system. The d'Alembert principle says that a constrained mechanical system performs motions in a way that the total virtual work of the constraint forces and torques is zero. For the case of rigid bodies it says that the virtual work of applied forces and torques dA equals the virtual work of current mass accelerations dB [28].

$$dA = dB \Rightarrow d(A - B) = 0 \tag{2.1.15}$$

From (2.1.15) with the mass matrix $\boldsymbol{M}$ follows:

$$d(A - B) = (q - \boldsymbol{M}\ddot{x})\,dx = 0 \tag{2.1.16}$$

The virtual displacements dx are not independent and thus need to be reduced to the number of generalised coordinates. In consequence, (2.1.15) can be rewritten by using the positional vector of generalised coordinates y and assuming a holonomic, unconstrained system. This leads to a system of ODE [9].

$$\boldsymbol{M}(y, t)\ddot{y} = q(y, \dot{y}, t) \tag{2.1.17}$$

For the formulation of the Lagrange equation of the first kind the positional vector x is used and the implicit constraints from (2.1.7) are applied by introducing the Lagrangian multipliers λ. Furthermore, the vector valued function $g(x)$ describing the constraints is added. The result is a system of differential algebraic equation (DAE) [28].

$$\boldsymbol{M}(x)\ddot{x} = q(x, \dot{x}, t) - \Gamma_x^T \lambda \tag{2.1.18}$$

$$0 = g(x) \tag{2.1.19}$$

2.1.2 Flexible Bodies

The equations of motion covered in the previous section apply to systems of rigid bodies. The simplifying assumption of rigid bodies is suitable for the calculation of the overall dynamic of bodies with negligible elastic deformations in a system, but it is not possible to obtain information on the deformation of specific bodies in the system. As the valves experience high dynamic loads, it is essential to introduce flexible bodies into the simulation model. Thus, this section expands the system equations for a system of rigid bodies to a system of equations also describing a system of flexible bodies.

2.1.2.1 Continuum Mechanics

The theory of elasticity describes the motion and internal forces of a solid, elastic body defined by a closed set of points. For the kinetic description physical properties are assigned to these points. A linear-elastic material behaviour is assumed.

The definition of points belonging to the considered continuum refers to a reference configuration. Most MBS related applications use the Lagrangian description of fields. In this notation, the continuum in its reference configuration is fixed to a coordinate system in a way that the coordinate grid performs the continuum's deformations. This is shown in the upper part of Fig. 2.1. In other words the coordinate reference is fixed to a material point of the deforming body. R_0 is the position vector of a point of the continuum in its reference configuration. The opposite option to describe the deformation of fields is the representation with Euler coordinates, shown in the bottom part of Fig. 2.1. Here, the grid lines representing the body deformation are applied to the deformed body. The coordinate reference is spatially-fixed.

To describe the deformation of the flexible body the deformation vector u is utilized. With the Lagrange representation this vector is defined [28]:

$$R = R(R_0, t) \tag{2.1.20}$$

$$u(R_0, t) = R(R_0, t) - R_0 \tag{2.1.21}$$

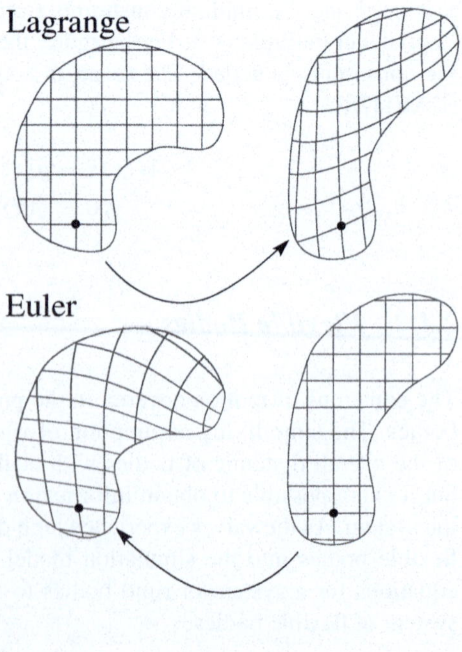

Fig. 2.1 Assignment of grid lines to a continuum in a reference configuration and a corresponding deformed configuration. Top: Deformation grid corresponds deformation representation in Lagrange coordinates. Bottom: Deformation grid corresponds deformation representation in Euler coordinates [39]

If u is the same for all points of a continuum the body performs a rigid body motion. If u is different for the points of one continuum, there is a deformation gradient between the points and therefore strains and stresses appear in the continuum.

Now, the linear-elastic strain tensor ϵ is introduced [28]. The indices i and j represent the three principal axis and therefore the position of the components in the strain tensor.

$$\epsilon_{ij} = \frac{1}{2}\left(\frac{\partial u_i}{\partial R_j} + \frac{\partial u_j}{\partial R_i}\right) \qquad i, j = 1, 2, 3 \tag{2.1.22}$$

The following description of the linear-elastic material law is based on the explanations in [28].

The material law is introduced to relate the strain tensor ϵ and stress tensor σ. C is the stiffness tensor describing the anisotropy of the elastic constants.

$$\boldsymbol{\sigma} = \boldsymbol{C}\boldsymbol{\epsilon} \tag{2.1.23}$$

For isotropic linear-elastic materials like steel, the stiffness tensor only depends on the Young's modulus E and the Poisson's ratio ν. The stiffness tensor can therefore be expressed by the two Lamé constants Λ and η that are defined by:

$$\Lambda = \frac{E\nu}{(1+\nu)(1-2\nu)} \tag{2.1.24}$$

$$\eta = \frac{E}{2(1+\nu)} \tag{2.1.25}$$

With that, the material law can be denoted as following:

$$\boldsymbol{\sigma} = \Lambda \mathrm{tr}(\epsilon)\boldsymbol{I} + 2\eta\epsilon \tag{2.1.26}$$

To obtain the equations of motion of the elastic body in Lagrangian representation, the momentum balance of a point P referring to the origin of the reference configuration is set up. From the Newton's equation in (2.1.1) follows that the momentum time derivative equals the external forces (2.1.27). b_0 defines the external surface forces on the infinitesimal surface element dA and d_0 defines the volume forces on the infinitesimal volume element dV. ρ_0 is the material's density.

In Lagrangian notation the momentum balance yields:

$$\frac{d}{dt}p^P = \frac{d}{dt}\int_V \rho_0(R_0)\dot{R}_0(R,t)\,dV = \int_V \rho_0(R)d_0(R,t)\,dV + \int_A b_0(R,t)\,dA \tag{2.1.27}$$

The first term of the right side of the equation describes the body forces and the second term describes the surface forces. If the reference system equals the inertial system, the deformation vector u with $u(R_0, t) = R(R_0, t) - R_0$ can be used to obtain the

Cauchy equations of motion. The Cauchy equations of motions are the local form of the Newton-Euler equations. The complete derivation is shown in [28, 39].

The d'Alembert principle already introduced in (2.1.15) is applied for flexible bodies, according to [39]. The virtual displacement and velocities are therefore δu and $\delta \dot{u}$. With respect to the virtual work of external forces, b_0 defines the external surface forces on the infinitesimal surface element A_0 and d_0 defines the volume forces on the infinitesimal volume element V_0.

$$\delta A = \int_{V_0} \delta u^T b_0 d V_0 + \int_{A_0} \delta u^T d_0 d A_0 \tag{2.1.28}$$

With the material density ρ the virtual work of mass accelerations is defined by:

$$\delta B = \int_{V_0} \delta u^T \rho \ddot{u} d V_0 \tag{2.1.29}$$

In addition, to the formulation for rigid bodies the virtual deformation work adds up for elastic bodies that is defined by:

$$\delta W = \int_{V_0} \delta \epsilon^T \sigma d V_0 \tag{2.1.30}$$

The virtual work of the external forces equals the sum of virtual work of mass acceleration and virtual work of deformation [28].

$$\delta A = \delta W + \delta B \tag{2.1.31}$$

From (2.1.27) follows (2.1.32). The left side of the equation follows from the virtual work of mass accelerations in (2.1.29). With D_0 describing the distribution of volume and surface forces resulting from (2.1.28) and the Cauchy stress distribution resulting from (2.1.30) follows:

$$\rho(R)\ddot{u}(R, t) = D_0(R, t) + \nabla \sigma \tag{2.1.32}$$

2.1.2.2 Flexible Bodies in the Multibody System

For the simulative analysis of flexible body deformations a wide range of FEM software is available. Usually these finite elements (FE) models have a large number of nodes and elements and therefore a high number of DOF. A coupling of such an FE model with the MBS approach would lead to immense calculation times. Hence, it is necessary to reduce the total number of DOF of the flexible body which can be done by various procedures, e.g. shown in [24, 35]. The FE model is therefore converted to a so-called superelement with reduced DOF.

Commercial MBS software like SIMPACK uses the component mode synthesis (CMS) method based on a procedure introduced by [6], originally invented for the substructuring of flexible structures.

For the application in MBS each flexible body is considered separately. So that also the reduction of DOF is performed for each flexible body of the system individually. The nodes of the FE model of the flexible body are subdivided into two groups. The nodes of the first group interact with other components of the MBS model through joints and constraints. Also external forces act on the flexible body on these nodes. These nodes yield the external DOF of the flexible body and are referred to as boundary points [6], interface nodes [7] or master nodes [24, 35].

The second group of nodes contains the nodes yielding the interior DOF [6]. No external forces or constraints act on these nodes. They are also referred to as slave nodes [24, 35].

The CMS consists of constraint modes and constrained normal modes [7]. The constraint modes represent the body's static shape for specific boundary conditions and are determined as follows: All interface nodes are fixed. Only one interface node per mode has one unconstrained DOF. A unity load is applied to the unfixed DOF. The result is the representation of one of the body's constraint modes. The procedure is repeated for each of the interface node's DOF to generate all constraint modes. These constraint modes are equivalent to Guyan modes [17]. For cases where only static deformation or low frequency modes of the body are important this approach leads to good results. The aim of the application of the unity load described above is the derivation of the transformation matrix $\boldsymbol{T}$. This matrix transforms displacement boundary conditions x_b on the interface nodes of the flexible body to the displacement vector that contains not only these boundary displacements, but also the interior displacements x_i. The internal displacements describe the displacement of each point of the flexible body when only the external deformation at an interface node is known.

By applying Hooke's force law, the transformation matrix can be determined. It is assumed that the stiffness matrix K is known. The indices ii refer to matrix entries for interior elements, bb refer to matrix entries for boundary elements and indices bi refer to coupling elements. The force vector includes the boundary forces F_b that are forces applied to the interface nodes and the interior forces F_i, however the interior forces are zero. The equation system is therefore defined by:

$$\begin{bmatrix} F_b \\ F_i \end{bmatrix} = \begin{bmatrix} F_b \\ 0 \end{bmatrix} = \begin{bmatrix} K_{bb} & K_{bi} \\ K_{bi} & K_{ii} \end{bmatrix} \begin{bmatrix} x_b \\ x_i \end{bmatrix} \tag{2.1.33}$$

This equation is solved for the unknown interior deformations of the flexible body x_i. This leads to the definition of the constraint modes ϕ_C.

$$x_i = -K_{ii}^{-1} K_{bi} x_b = \phi_C x_b \tag{2.1.34}$$

Hence, the transformation matrix $\boldsymbol{T}$ results in (2.1.35).

$$T = \begin{bmatrix} I \\ -K_{ii}^{-1} K_{bi} \end{bmatrix} \tag{2.1.35}$$

For the displacement $\bar{x}$ follows with this transformation matrix:

$$\begin{bmatrix} x_b \\ x_i \end{bmatrix} = T x_b \tag{2.1.36}$$

To be able to consider higher frequency contents of the body deformation, normal modes have to be determined. They are defined as the body's normal modes with all interface nodes being fully constrained, which leads to the term of constrained normal modes. Thus, they can be obtained from the eigenvalue problem shown in (2.1.37) and introduced by [14], whereby the eigenvectors ϕ_j form the columns of the normal mode matrix ϕ_N. The corresponding eigenfrequency for the eigenvector is ω_j.

$$(-\omega_j^2 m_{ii} + k_{ii})\phi_j = 0 \tag{2.1.37}$$

The transformation matrix resulting from this now includes the constraint modes and the normal modes [6].

$$T_{CMS} = \begin{pmatrix} I & 0 \\ \phi_C & \phi_N \end{pmatrix} \tag{2.1.38}$$

The matrices defining the FE model are now transformed to matrices describing the superelement. The equations are shown in (2.1.39) to (2.1.42). The term superelement is derived from the original application of the reduction technique, the substructuring of an FE model. In the present case, the term superelement refers to the reduced flexible body.

$$M_{SE} = T_{CMS}^T M_{FEM} T_{CMS} \tag{2.1.39}$$

$$D_{SE} = T_{CMS}^T D_{FEM} T_{CMS} \tag{2.1.40}$$

$$K_{SE} = T_{CMS}^T K_{FEM} T_{CMS} \tag{2.1.41}$$

$$F_{SE} = T_{CMS}^T F_{FEM} \tag{2.1.42}$$

The modal approach, by introducing the reduced superelement matrices and vectors, is used to describe small elastic deformations by superposition of interface modes and eigenmodes. The evaluation of the superelement's eigenmodes is similar to (2.1.37), now with the superelement matrices:

$$(K_{SE} - \omega^2 M_{SE})\phi_i(R_0) = 0 \tag{2.1.43}$$

Interface modes can be added to obtain additional information about the body's deformation close to interface nodes. They can be expressed as frequency response modes (FRM) or as inertia relief modes. For the determination of FRM, the modes are defined by (2.1.44) [14]. The FRM is the body's response to a harmonically time-dependent unit load p_{SE} applied to the flexible body [8]. Ω is the excitation frequency.

$$(\boldsymbol{K}_{SE} - \Omega^2 \boldsymbol{M}_{SE})\phi_i(R_0) = p_{SE} \tag{2.1.44}$$

With decreasing main frequency, the modes of (2.1.44) represent the quasi-static body displacement due to a unit load applied to one of the interface nodes. The forces due to the constant acceleration caused by the unit load are balanced due to inertial forces applied to the structure in a way that (2.1.45) applies, with $F_{inertial}$ describing the balancing inertial forces and $a_{inertial}$ describing the acceleration field resulting from the quasi-static displacement.

$$F_{inertial} + \int_V a_{inertial} \rho dV = 0 \tag{2.1.45}$$

The results of this configuration are the inertia relief modes (IRM). The results of the IRM or FRM analysis don't deliver additional information on flexible body deformation for the case that all accessible eigenmodes are used. A reason for not using all accessible eigenmodes is the reduction of simulation time for a decreasing number of considered eigenmodes as pointed out by [45]. All unconsidered modes are transferred to FRM or IRM modes [7].

For solving the resulting partial differential equations, the approaches of Ritz and Galerkin are applied. These approaches are based on the principle of separation of variables. Therefore, this method divides the flexible body deformation $u(R_0, t)$ in a time-dependent part and a time-invariant part. Time-invariant eigenfunctions build the initial functions. Infinite series of these initial functions are able to form any function. Weighting factors of the initial functions are the time-variant modal coordinates $s_{flx,i}$. These modal coordinates are determined during simulation. For the finite number of considered modes N_R (2.1.46) describes the deformation [39].

Another approach using Ritz' method is the nodal approach instead of the modal approach. The nodal approach uses the interpolation functions given by the FE model of the flexible body as shape functions. This approach delivers high accuracy results of the flexible body deformation [41].

However, this approach goes along with a high computational effort and is therefore not used for the analyses in this thesis.

$$u(R_0, t) = \sum_{i=1}^{N_R} \phi_i(R_0) s_{flx,i}(t) \tag{2.1.46}$$

2.1.2.3 Modal Stresses

There are two approaches to determine the stresses and strains of the flexible body. The one approach requires the transformation of the determined component loads to the FE model to determine the stresses of the component. Some software packages provide co-simulation options of MBS and the corresponding FE model. Apart from the additional data transfer of big data sets, the main disadvantage of this approach is that the FE simulation is based on a quasi-static computation. As a result, the effects of vibrations are not considered. For a component with high dynamic loads like valve train components these effects of vibration are the main load. As a consequence, this approach leads to big errors in the load determination.

Therefore, the determination of stresses and strains of the flexible bodies in this thesis follows the modal stress approach presented in [8, 13]. Schwertassek et al. [40] show a good quality of the resulting stresses when the dynamic boundary conditions of the system are considered properly. This can be reached by considering flexible and static body modes.

The modal stress approach follows a similar scheme as the modal approach utilised in the CMS shown in the previous section. Each mode contains a specific mode shape and also a specific stress distribution, the so-called modal stresses. The stress state of the component is therefore a linear combination of the weighted contributions of all the modes considered for the component. Equivalent to the modal deformations the weighting of the modal stress contribution depends on the modal coordinates. The transient stress s_σ of the component is therefore defined as:

$$s_\sigma(R_0, t) = \sum_{i=1}^{N_R} \sigma_i(R_0) s_{flx,i}(t) \tag{2.1.47}$$

This method requires no additional FE analysis, the modal stresses of each mode and the corresponding modal coordinates can directly be analysed within the MBS tool. Afterwards, the stresses are transferred to further software for a subsequent fatigue assessment analysis.

2.1.2.4 Damping

The term damping describes the dissipation of energy in a mechanical system that is capable of oscillation. That definition applies to an MBS system. The damping effects that occur at the flexible bodies have not been considered yet, because the damping is applied after the modal reduction of the flexible body during the MBS.

The MBS software SIMPACK uses a damping model, defined in (2.1.48) that adds up the velocity depending viscous damping and and amplitude depending structural damping in order to evaluate the overall flexible body damping effects [7]. a_{damp} is the damping scaling factor. This factor is determined by SIMPACK and results from the equivalent summation of structural damping and modal damping of the system.

The viscous damping matrix $\boldsymbol{D}_{v,modal}$ is computed from the substructure's viscous damping matrix and the equivalent modal viscous damping matrix $\boldsymbol{D}_{eq,modal}$ is computed from the structural damping matrix of the substructure. Therefore, the structural damping is transformed to an equivalent viscous damping. The equation of flexible body damping holds:

$$\boldsymbol{D}_{flex} = \boldsymbol{D}_{modal} + a_{damp}(\boldsymbol{D}_{v,modal} + \boldsymbol{D}_{eq,modal}) \tag{2.1.48}$$

The terms of this equation are explained in the following.

The matrices $\boldsymbol{D}_{v,modal}$ and $\boldsymbol{D}_{eq,modal}$ consider the material dependent Rayleigh damping and local damping effects if present. The structural damping values are transformed into equivalent viscous damping values and added to the viscous damping values.

The steps necessary to transform the substructure structural damping matrix to the equivalent viscous damping matrix are outlined in the following.

Firstly, the modal structural damping matrix $\boldsymbol{D}_{struc,modal}$is obtained from the substructures damping matrix $\boldsymbol{D}_{struc}$ with the mode shape matrix $\boldsymbol{\Phi}$ by a similarity transformation:

$$\boldsymbol{D}_{struc,modal} = \boldsymbol{\Phi}^T \boldsymbol{D}_{struc} \boldsymbol{\Phi} \tag{2.1.49}$$

Secondly, the modal structural matrix is transformed by the transformation matrix $\boldsymbol{T}_{Damp}$, where f_i is the frequency of the mode i:

$$\boldsymbol{T}_{Damp} = \mathrm{diag}\left(\frac{1}{\sqrt{f_i}}\right) \tag{2.1.50}$$

$$\boldsymbol{D}_{eq,modal} = \boldsymbol{T}_{Damp}^T \boldsymbol{D}_{struc,modal} \boldsymbol{T}_{Damp} \tag{2.1.51}$$

The modal damping matrix depends on the damping type selected in the MBS software. Either a critical damping or a stiffness proportional damping can be selected. α_i is thus the critical damping factor of mode i, whereas β_i is the stiffness proportional damping factor of mode i. The modal damping matrix (2.1.53) consists of the modal damping values d_i:

$$d_i = \begin{cases} 2\alpha_i \sqrt{k_i m_i} \\ \beta_i k_i \end{cases} \tag{2.1.52}$$

This leads to the modal damping matrix:

$$\boldsymbol{D}_{modal} = \mathrm{diag}(d_i) \tag{2.1.53}$$

2.1.2.5 Kinematic Description of a MBS

The following sections show the integration of a flexible modally reduced body into a MBS model. This section starts with the kinematic description of body motion and deformation. The next section shows the kinetic description of the flexible body and the derivation of the equations of motion.

The kinematic description uses different coordinate systems to define the deformation of a flexible body system. The coordinate systems are defined by their origin O^i and their unit vectors e^i. The inertial coordinate system is identified by $\{O^I, e^I\}$.

The floating frame of reference is widely used to consider flexible body deformation within a MBS model. It includes a subdivision of rigid body deformation and elastic deformation of flexible bodies. The elastic deformations are described with respect to a body reference configuration. The elastic deformations are assumed to be small so that the equations of motion can be linearised with respect to the deformation coordinates, while the overall reference motion may be large [39, 42].

Figure 2.2 shows the configuration of the reference body configuration in dark grey with its reference coordinate system $O^i, e^i\}$. The light grey body represents the deformed state of this reference configuration. The orientation of the reference system $\{O^i, e^i\}$ with respect to the inertial system $\{O^I, e^I\}$ is described by the rotational matrix $\boldsymbol{A}$ (2.1.54) that is the transformation matrix of the reference body frame

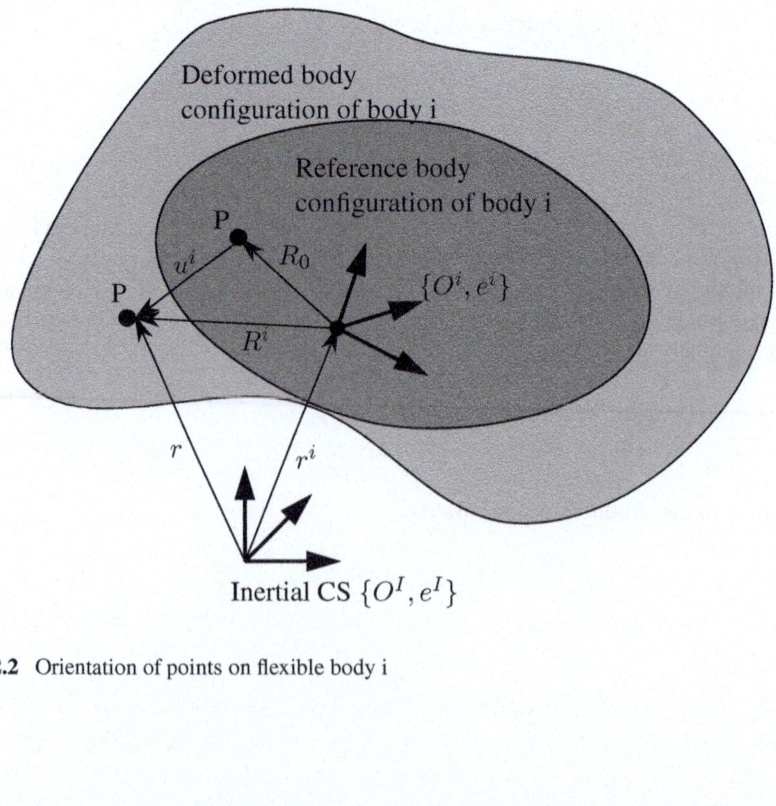

Fig. 2.2 Orientation of points on flexible body i

with respect to the inertial coordinate system. With this matrix, the angular velocity $\tilde{\omega}$ with respect to the reference frame can be determined.

$$e^i(t) = \boldsymbol{A}^i(t)e^I \tag{2.1.54}$$

$$\tilde{\omega}^i = \boldsymbol{A}^i \dot{\boldsymbol{A}}^{i^T} \tag{2.1.55}$$

With (2.1.55) the position of each point on the reference body relative to the inertial system can be derived. Therefore, R_0 is the location of point P in the body's reference configuration.

$$r^I(R_0, t) = \boldsymbol{A}^{i^T}(t) r(R_0, t) \tag{2.1.56}$$

From that, the position (2.1.56), velocity and acceleration for an arbitrary point P on the flexible body can be derived [39]:

$$r(R_0, t) = r^i(t) + R_0 + u^i(R_0, t) \tag{2.1.57}$$

$$v(R_0, t) = v^i(t) + \dot{u}^i(R_0, t) + \tilde{\omega}^i(t)(R_0 + u^i(R_0, t)) \tag{2.1.58}$$

$$a(R_0, t) = a^i(t) + \ddot{u}^i(R_0, t) + 2\tilde{\omega}^i(t)\dot{u}^i(R_0, t) + (\dot{\tilde{\omega}}(t) + \tilde{\omega}(t)\tilde{\omega}(t))(R_0 + u^i(R_0, t)) \tag{2.1.59}$$

It is now possible to replace the deformation vector u in the kinematic equation (2.1.57) by the modal approach defined in (2.1.46). The resulting equation is then inserted into (2.1.56), which leads to [7]:

$$r^I(R_0, t) = \boldsymbol{A}^{i^T}(t)(r^i(t) + R_0 + \sum_{i=1}^{N_c} \phi_i(R_0) s_{flx,i}(t)) \tag{2.1.60}$$

The equations of motion can be written in state space form. This form is introduced in the following section. From the kinematic perspective, this form requires state space vectors. These vectors contain all information regarding position and velocity of the flexible body. The vectors z_I and z_{II} contain the flexible body states s_{flx} [39]. These flexible states contain the positional and velocity state vectors.

$$z_I = \begin{bmatrix} r^i(t) \\ \alpha^i(t) \\ s^i_{flex}(t) \end{bmatrix} z_{II} = \begin{bmatrix} v^i(t) \\ \omega^i(t) \\ \dot{s}^i_{flex}(t) \end{bmatrix} \tag{2.1.61}$$

2.1.2.6 Equations of Motion for Flexible Multibody Systems

To derive the equations of motions from the kinematic description in (2.1.60) again, d'Alembert principle from (2.1.15) applies and is denoted in (2.1.62).

$$dA = dW_S + dB \tag{2.1.62}$$

If the reference coordinate system is the inertial coordinate system the virtual displacements can be defined as follows:

$$\delta r = \delta u \tag{2.1.63}$$

Now, the integrals of virtual work can be determined. For a flexible body the virtual strain work W_s has to be considered and is defined in the second term on the right side of (2.1.64). The left side of the equation describes the work of the volume forces and surface forces, the first term on the right side describes the virtual work of the mass acceleration forces.

$$\int_V \delta u^T d_0 \, dV + \int_A \delta u^T b_0 \, dA = \int_V \rho \delta u^T \ddot{u} \, dV + \int_V \delta \epsilon^T \sigma \, dV \tag{2.1.64}$$

The matrix of generalised masses results from the mass acceleration forces in the previous equation [39]. On its main diagonal, the mass matrix contains submatrices referring to translations and rotations of the reference system (M_{rr} and M_{tt}). The submatrix M_{ee} refers to the deformations of bodies. The mass matrix contains not only diagonal entries but also contains coupling elements describing the rigid-elastic contribution and the translational-rotational contribution.

$$\boldsymbol{M^i} = \begin{bmatrix} M^i_{tt} & & \text{sym.} \\ M^i_{rt} & M^i_{rr} & \\ M^i_{et} & M^i_{rr} & M^i_{ee} \end{bmatrix} \tag{2.1.65}$$

Similar to the matrix of generalised masses, the other terms of (2.1.64) hold the vector of generalised forces. The volume forces and surface forces are on the left side of the equation, while the second term on the right side of the equation holds the inner forces. These forces are either part of the applied forces or of the constraint forces of the system.

Because the floating frame of reference is used to describe the flexible body deformation, gyroscopic terms have to be considered on the side of applied forces in the system's equations of motion. Besides the gyroscopic terms, the function of applied force F contains the discrete joint forces as well as the continuous imposed forces.

The section describing the kinematic of the multibody system already introduced the vectors z_I and z_{II} describing the positions and velocities of the mechanical system. Now the additional matrix of coefficients $\boldsymbol{Z}$ is introduced. This matrix relates the derivatives of the positions to the velocities of the system.

With the position and velocity variables stated in (2.1.61) the system's equations for the MBS considering flexible and rigid body deformation can be denoted similar to the system equations for an all rigid MBS, (2.1.18).

$$\dot{z}_I = \boldsymbol{Z} z_{II} \tag{2.1.66}$$

$$\boldsymbol{M} \dot{z}_{II} - \Gamma^T \lambda = F \tag{2.1.67}$$

$$g(z_I, t) = 0 \tag{2.1.68}$$

A further option to describe this system of differential equations is the state space representation. This representation consists of two equations per system, the state space differential equation and the output equation. The system is therefore described by a set of inputs, outputs and system states. A second order differential equation is transformed to a first order differential equation. Therefore, the MBS simulation as well as several further simulation tools, use the state space form.

The variables holding positions z_I, velocities z_{II}, the constraint forces and torques λ, and also the dynamic states c holding the dynamic information of force elements and algebraic states s can be summarised in the state vector x_{State} [7]

$$x_{State}(t) = \begin{bmatrix} z_I \\ c \\ z_{II} \\ \lambda \\ s \end{bmatrix} \tag{2.1.69}$$

With this state space vector, the dynamic MBS system can be described in the state space form:

$$\dot{x}_{State}(t) = \boldsymbol{A}_{Sys} x_{State}(t) + \boldsymbol{B}_{Inp} u_{Inp}(t) \tag{2.1.70}$$

$$y_{out}(t) = \boldsymbol{C}_{Sys} x_{State}(t) + \boldsymbol{D}_{Inp} u_{Inp}(t) \tag{2.1.71}$$

Beside the state vector there is the vector of inputs to the system u_{Inp}. The vector y_{out} holds the outputs of the dynamic system.

The matrix $\boldsymbol{A}_{Sys}$ is the system matrix. It holds the coefficients of the state variables in x_{State}. The matrix $\boldsymbol{B}_{Inp}$ defines the input matrix that describes the influence of the inputs on the state variables. The output matrix $\boldsymbol{C}_{Sys}$ multiplied with the state variables describes the dynamic effect of the state variables on the output variables. The static influence of the output variables is a linear combination of the inputs. These linear combinations are denoted by the feedthrough matrix $\boldsymbol{D}_{inp}$.

2.1.3 Linearised Equations of Motion

For certain applications, the linearisation of the equations of motion is used. According to [7], the eigenvalue solver needs a linearised version of the in general non-linear equations of motion stated in Sects. 2.1.1 and 2.1.2. The linear equations in general form stated in (2.1.70) with the state vector from (2.1.69) describe the eigenbehaviour of the MBS system and also the response to excitation. The matrix A_{Sys} is the only required component for the eigenvalue analysis because it describes the system. For an accurate linearisation, the following two preconditions must be fulfilled: The system must be in or close to the equilibrium state and no discontinuities are allowed to occur in the system at the point of linearisation. The influences of equilibrium state and discontinuities on the linearisation are explained in the following.

Equilibrium State: The system states of velocity, acceleration and joint and constraint state derivatives should be as close to zero as possible. The reason is that the Taylor approximation utilised to describe the transfer function is truncated after the second term and therefore does not include effects of velocities and accelerations in the system (see also (2.1.74)).

Discontinuities: The eigenvalue solver has no discontinuity handling (see also Sect. 2.1.5) and would use the mean value between the two iteration steps where the discontinuity occurs. This is unrealistic for most cases. Discontinuities are e.g. jumps in the force characteristic or discrete changes in contact states (e.g. from 'in contact' to 'not in contact'). This precondition applies to all elements using root functions.

The joint connection between flexible bodies is considered regarding their damping and stiffnesses. According to [29], the partial derivatives are directly added to the modal stiffness matrix at the positions of the connected nodes of the flexible bodies.

The previously introduced state space representation of the dynamic system is typically used to perform the complex modal analysis, taking e.g. viscous damping into account [30]. As mentioned before, the only component of the linear system (2.1.70) necessary for the description of the system's eigenbehaviour is the matrix $\boldsymbol{A}_{Sys}$, so that the eigenvalues (2.1.72) can be solved. $\boldsymbol{I}$ is the identity matrix, e_i are the i unit vectors, ζ_i are the eigenvalues corresponding to e_i .

$$\boldsymbol{A}_{Sys} e_i = \zeta_i \boldsymbol{I} e_i \tag{2.1.72}$$

An approximation function $f(x_{State})$ is used to describe the non-linear system's derivative of state vector (2.1.73). The function $f(x_{State})$ is estimated by a Taylor series expansion.

$$\dot{x}_{State} = f(x_{State}) \tag{2.1.73}$$

$$\dot{x}_{State} = f(x_{State,0}) + \left.\frac{\partial f}{\partial x_{State}}\right|_{x_{State,0}} (x_{State} - x_{State,0}) + \ldots \tag{2.1.74}$$

During the linearisation for the linearisation state $x_{State,0}$ the Taylor series expansion is truncated after the second term and therefore approximates $\dot{x}_{State}$ as follows:

$$\dot{x}_{State} \approx f(x_{State,0}) + \left.\frac{\partial f}{\partial x_{State}}\right|_{x_{State,0}} (x_{State} - x_{State,0}) \tag{2.1.75}$$

The infinitesimal difference quotients occurring in (2.1.75) are approximated by finite differences (2.1.76) where the increment size Δx_{State} depends on the machine precision, a user defined scaling factor and the state values $x_{i,0}$.

$$\frac{\partial f_j}{\partial x_{State,i}} \approx \frac{\Delta f_j}{\Delta x_{State,i}} \tag{2.1.76}$$

The system matrix $\boldsymbol{A}_{Sys}$ is described by the remaining difference quotient:

$$\boldsymbol{A}_{Sys} = \left.\frac{\Delta f}{\Delta x_{State}}\right|_{x_{State,0}} \tag{2.1.77}$$

The application of this method to the modal analysis of an internal combustion engine in [29] shows, that the deviation between the results of the linearised equations of motion and the FE approach is smaller than 10%. Mostly the deviations are even smaller than 1%.

2.1.4 Contact Formalisms

The contact pairing of bodies in an MBS system results in forces on both contacting bodies. With respect to the equations of motion, this might either be an applied external force or a constraint force [2]. In many cases, a linear behaviour can be assumed if the contact appears between rigid bodies or if the elastic deformation of a flexible body in contact is negligible. Thus, unilateral spring-damper force elements are used for contact modelling. Hooke's law applies here. Moreover a root state depending formulation enables the detection of whether the bodies are in contact or not and therefore determine model discontinuities (see also Sect. 2.1.5.2). The solver must have a root state detection and root state handling in that case.

The subsequent subsections explain the approaches on the analysis of non-linear contact behaviour in MBS models. The non-linear contact behaviour is considered e.g. for elastic or non-elastic contact patches with multiple contact points or for non-negligible elastic deformation in the contact region.

2.1.4.1 Polygonal Contact Model

The polygonal contact model (PCM) contact formalism introduced by [20] enables the sufficient approximation of contact force for complete contact patches of two bodies even with multiple areas of contact.

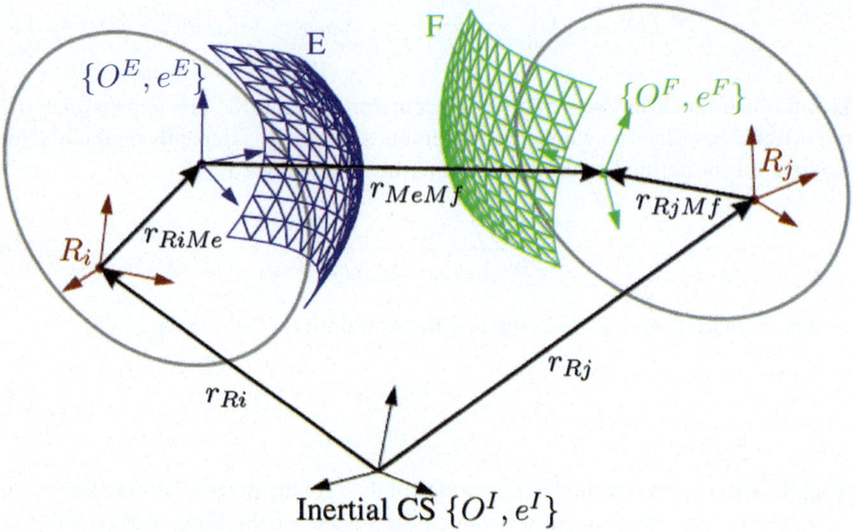

Fig. 2.3 Reference frames of PCM contact [20]

A discretisation of the contact patch area enables the calculation of the contact force of each polygon separately. These forces are summed up afterwards to obtain the resulting contact forces acting on the bodies. Hence, the integration of PCM contact into an MBS system is simple because the PCM acts like an ordinary force element.

The contact patches of both contacting bodies with free form surfaces are approximated by polygon surfaces. The contact dynamic is described by the elastic foundation model. This model postulates a thin elastic layer with a thickness b on rigid bodies with an elastic modulus K resulting from the material parameters:

$$K = \frac{1-\nu}{(1+\nu)(1-2\nu)} \cdot E \tag{2.1.78}$$

The approach is also applicable for elastic bodies when the layer stiffness $cl = \frac{K}{b}$ can be determined by qualified estimation. Then the normal contact force $F_{c,N}$ for each contact element with the contact area A_c and the normal displacement u_N is defined by (2.1.79).

$$F_{c,N} = c_l \cdot A_c \cdot u_N \tag{2.1.79}$$

Figure 2.3 shows the definition of the reference frames for this contact. The PCM contact element uses the reference frames $\{O^E, e^E\}$, $\{O^F, e^F\}$ for each of the contacting bodies that represent the surfaces E and F in contact.

Firstly, a collision detection is performed. For an efficient implementation, the detection algorithm is based on bounding volume hierarchies described in [46].

The bounding volumes are cuboids and represent a hull of contiguous parts of the surface. The root element of this hierarchical structure includes all surfaces of the body. With each additional level in the binary tree hierarchy, the elements are divided into two child boundary volumes. Thus, the collision analysis does not have to be performed for each possible combination of contact and target surface polygons but only for the ones with colliding bounding volumes.

For colliding polygons, the analysis of the contact penetration area is performed afterwards. Hippmann [20] outlines details on the determination of the contact penetration geometry. During this process, the normal vector n_{pcm} of the master surfaces of each contact element is analysed. The overall contact force per contact element is the sum of tangential and normal force. The latter is composed of a stiffness depending part and a damping depending part defined by (2.1.80). The damping depends on the normal share v_N of the overall body velocity v_{pcm} and the damping factor d_l.

$$F_{d,N} = d_l \cdot A_c \cdot v_N \tag{2.1.80}$$

The normal force F_N can be obtained now.

$$F_N = F_{c,N} + F_{d,N} \tag{2.1.81}$$

The tangential force F_T depends on the relative tangential velocity v_T and the friction. The PCM contact uses the Coulomb regularised friction model [33]. Thus, a small velocity limit v_ϵ is set. The friction force is faded out below this limit. μ is the friction coefficient.

$$v_T = |\mathrm{v}_{pcm} - v_N| \tag{2.1.82}$$

$$F_T = \begin{cases} \mu \cdot F_N, & v_T \geq v_\epsilon \\ \mu \cdot F_N \cdot \frac{v_T}{v_\epsilon}\left(2 - \frac{v_T}{v_\epsilon}\right), & v_T < v_\epsilon \end{cases} \tag{2.1.83}$$

The total force vector of each contact element is obtained by the sum of the projection of the normal force component along its normal vector and the tangential force component (2.1.84) counteracting on the direction of the tangential velocity vector v_T. The resulting torque is determined by the force acting on body E at the reference frame $\{O^F, e^F\}$. The specific contact position is represented by C_k. The sum of all forces F_{pcm} and torques M_{pcm} of the k contact elements results in the total force F_E and torque M_E of the contact.

$$F_{pcm} = F_N \cdot n_{pcm} + F_T \cdot \frac{\mathrm{v}_T}{v_T} \tag{2.1.84}$$

$$F_E = \sum_k F_{pcm}, \qquad k = 1, ..., n \tag{2.1.85}$$

$$M_{pcm} = (\mathrm{r}_{M_e C_k} - \mathrm{r}_{M_e M_f}) \times F_{pcm} \tag{2.1.86}$$

$$M_E = \sum_k M_{pcm}, \qquad k = 1, ..., n \tag{2.1.87}$$

2.1.4.2 Hydrodynamic Bearings

The formalism of EHD simulation is mainly used for the modelling of bearing support on elastic journal structures. Because the elastic deformations of journal and housing affect the oil film distribution and therefore the bearing dynamic characteristic significantly, it is important to access the hydrodynamic in bearings accurately.

The Reynolds equation describes the pressure distribution of the oil film in the bearing. This equation is derived from the Navier-Stokes equation for incompressible flow [10]. According to [25] the Reynolds equation can be written as (2.1.88) with the lubrication gap function h. P_{oil} is the oil film pressure, μ_{vis} is the lubricant viscosity. The journal radius is R_{jour}

$$\frac{\partial}{\partial z}\left(h^3 \frac{\partial P_{oil}}{\partial z}\right) + \frac{\partial}{\partial x}\left(h^3 \frac{\partial P_{oil}}{\partial x}\right) = 6 \cdot \mu_{vis} \cdot \omega \cdot R_{jour} \frac{\partial h}{\partial x} + 12 \cdot \mu_{vis} \cdot h \tag{2.1.88}$$

This equation is valid for both smooth or rough surfaces. With boundary conditions of the hydrodynamic pressure field and cavitation boundary conditions and a discretisation of the equation (2.1.88), it can be solved numerically. The impact of boundary conditions is visualized by Fig. 2.4.

The Reynolds equation can lead to negative pressures if Dirichlet boundary conditions are applied (Sommerfeld solution). This is a non-physical solution because lubricant can transmit no relevant tension forces. If the pressure curve is cut at zero, the resulting solution of Reynolds equation is the Gümbel solution. It has the disadvantage of a discontinuity in the pressure curve. The additional application of Neumann boundary conditions leads to the Reynolds solution. The boundary condition ensures the avoidance of discontinuities by stipulating a horizontal tangent curve on the pressure curve in the cavitation area [22].

The surface quality has no relevant influence on the pressure in the gap if the gap width is in a higher range of size than the surface roughness. However, if the size range is similar, the influence of the surface roughness on the pressure distribution must be considered.

With respect to roughness peaks of both the journal and the shell surface of the EHD contact, the local gap function h is the sum of the nominal film thickness $h*$ and the roughness amplitudes $\delta_{ref} + \delta_{opp} = \delta_{r,o}$ [3]

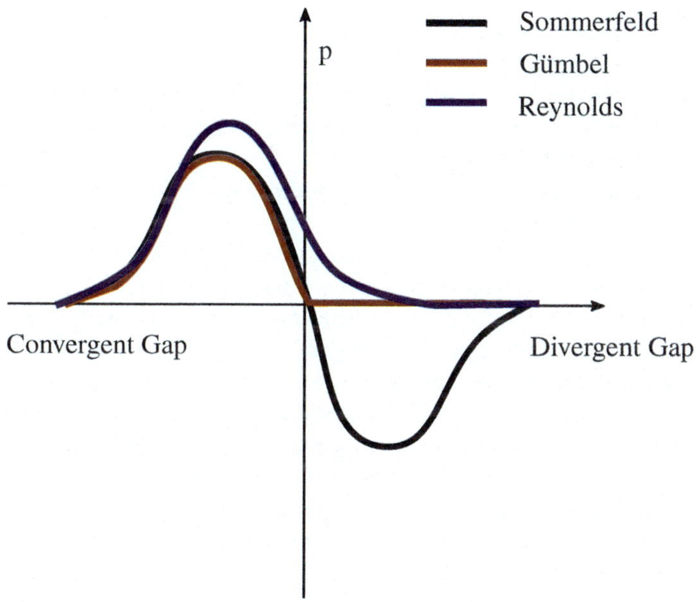

Fig. 2.4 Pressure curve for Sommerfeld, Gümbel and Reynolds solution of Reynolds equation

$$h = h * + \delta_{ref} + \delta_{opp} \tag{2.1.89}$$

With the assumption of a distribution function $f(\delta)$ for the roughness amplitudes e.g. the Gaussian distribution, the standard deviation $\sigma_{r,o}$ for the reference and opposed surface roughness can be expressed as:

$$\sigma_{r,o} = \sqrt{\int_{-\infty}^{\infty} \delta_{r,o} f(\delta_{r,o}) d\delta_{r,o}} \tag{2.1.90}$$

Therefore, the combined roughness density function parameter σ_δ is derived by:

$$\sigma_\delta = \sqrt{\sigma_{ref}^2 + \sigma_{opp}^2} \tag{2.1.91}$$

The Greenwood-Tripp model [15] describes the solid-state contact. With a Gaussian distribution of the surface roughness, the contact pressure p_{GT} is defined with the contact stiffness C_{GT} as:

$$p_{GT} = C_{GT} \cdot 4.4086 \cdot 10^{-5} \cdot (4 - \frac{h}{\sigma_\delta})^{6.804} \tag{2.1.92}$$

2.1.5 Solving

So far, this chapter introduced the methods and their physical background of modelling a mechanical system consisting of multiple bodies. The last step of the simulation is the solving of this system.

This section gives an overview over the methods used to solve ODEs or DAEs in the context of multibody systems.

Before solving the actual time-dependent system equations, the equilibrium state of the system can be calculated, although it is not obligatory. This applies to systems of ODEs and also to systems of DAEs. The term equilibrium state refers to a system with all accelerations and velocities being zero. In reality, states with very small but non-zero accelerations and velocities are also referred to as equilibrium states. A system in its equilibrium state is stabilised and thus has no or only very small vibrations left in body displacement, deformation or force states. The equilibrium state delivers initial values for the subsequent solving process.

Solving the MBS model requires a time integration solver. Time integration solvers like differential algebraic system solver (DASSL) are mainly based on Newton's method or backwards differentiation formula (BDF). Furthermore, commercial software like SIMPACK provides the option of including a discontinuity handling.

2.1.5.1 Equilibrium State Solving

The procedures for the evaluation of the equilibrium state of a MBS are described by [11].

In its equilibrium state the system does not move. Therefore, the velocities and accelerations are approximately zero. There are two methods to transform the system into an equilibrium state: The first method is the adjustment of the nominal forces, also referred to as preloads, in a way that the system is transformed into an equilibrium state for a given position. Therefore, the solution of a system of linear equations is required.

Firstly, the force distribution matrix $\boldsymbol{N}$ is introduced. This matrix relates the derivatives of the applied forces f_a in the MBS to the derivatives of the force vector f_e that assembles the contributing components of every interconnection between the parts of the MBS.

$$\boldsymbol{N} = \left(\frac{\partial f_a}{\partial f_{e_i}}\right) \qquad i = 1, \ldots, n_e \tag{2.1.93}$$

In order to obtain the unknowns consisting of the components of the interaction force vector f_e and the constraint force vector λ, the given position p_N has to satisfy the position constraints (2.1.94).

$$g(p_N) = 0 \tag{2.1.94}$$

Therefore, the system's force balance holds:

$$0 = f_a(f_e, p_N) - \mathbf{\Gamma}(p_N)^T \lambda \tag{2.1.95}$$

With the introduction of the force distribution matrix and the assumption of a linear relation of the applied forces f_a to the interaction forces f_e, the linear equation for the determination of the preloads can be obtained:

$$(\boldsymbol{N}(p_N), -\mathbf{\Gamma}(p_N)^T) \begin{pmatrix} f_e \\ \lambda \end{pmatrix} = -f_a(0, p_N) \tag{2.1.96}$$

This method mainly provides the information of a force equilibrium. Because of its linear approach, it does not deliver all initial conditions for the non-linear time integration. Therefore, another method delivering these information is introduced: The second method of equilibrium calculation is the adjustment of the position for given nominal forces in order to obtain the equilibrium state. This task leads to a system $\boldsymbol{F}$ of non-linear equations (2.1.97).

$$\boldsymbol{F}(x) = \begin{pmatrix} f_a(p, 0) - \mathbf{\Gamma}^T(p)\lambda \\ g(p) \end{pmatrix} = 0, \qquad x^T = (p^T, \lambda^T) \tag{2.1.97}$$

The solution of this equation (2.1.97) system holds important information on the mechanical system and delivers consistent initial values for the time integration of the equations of motion. The solution of this non-linear equation system also provides a point for linearisation in order to perform linear system analysis like the modal analysis in Sect. 2.1.3. Therefore, the non-linear approach delivers more information than the preload approach. However, it is more costly regarding simulation time.

2.1.5.2 Solving Non-linear Equations

This section introduces the solver used to solve MBS systems. MBS systems are often stiff problems due to e.g. high damping force elements or the implementation of springs [1]. The term stiff problem refers to a system of differential equations with numerical instabilities. These numerical instabilities lead to a failure of some numerical methods, explicit solvers in particular, or very small step sizes.

The SIMPACK solver as well as several further MBS solvers are based on the DASSL algorithm. To understand the DASSL, it is necessary to introduce Newton's method and the BDF. Therefore, the section starts by introducing these two methods before finally introducing the DASSL algorithm.

This section presumes the general solvability of the non-linear equation system. [11] gives an overview over details on criteria for solvability of a system. This section uses a general nomenclature as the methods refer not only to mechanical systems. Functions are described by $\boldsymbol{F}(x^k)$ with the function variable x.

Newton's Method

Newton's Method is a common approximating method to solve non-linear equations numerically. Its definition can be written as (2.1.98), where $\boldsymbol{F}'(x^n)$ is the Jacobian matrix. n is the current solving step.

$$x^{n+1} = x^n - \frac{\boldsymbol{F}(x^n)}{\boldsymbol{F}'(x^n)} \tag{2.1.98}$$

This equation can be transformed to the form that is solved numerically with the step size Δx [11].

$$\boldsymbol{F}'(x^n)\Delta x^n = -\boldsymbol{F}(x^n) \tag{2.1.99}$$

$$x^{n+1} = x^n + \Delta x^n \tag{2.1.100}$$

Newton's method requires the evaluation of the Jacobian in each iteration step which is very costly regarding calculation time. Therefore, algorithms reevaluating Jacobians from former steps are established and referred to as "simplified Newton's method".

Usually, finite differences are enabled to approximate the Jacobian matrix with the increment τ and the unit vector e with $||e|| = 1$

$$\frac{\partial F}{\partial x} \approx \frac{\boldsymbol{F}(x + \tau e) - \boldsymbol{F}(x)}{\tau} \tag{2.1.101}$$

The increment τ must be small enough to ensure the negligibility of the approximation error.

Backwards Differentiation Formula

The BDF is a class of multistep formalism. The theoretical background for linear multistep solvers is introduced by [4, 11]. The linear multistep algorithm's application is the approximation of solutions of ODEs. The implicit BDF is applied within a predictor-corrector scheme. Within a predictor-corrector scheme, the implicit value x_{n+1} is approximated by an explicit algorithm. Afterwards, the value is applied in an implicit corrector-step (e.g. by applying Newton's method) to obtain an adjusted value of x_{n+1}. This scheme is passed multiple times, until the difference between the predicted value of x_{n+1} and the corrected value of x_{n+1} is smaller than the accepted error tolerance.

For the BDF, the polynomial Π_k interpolates k points and can be expressed in Lagrangian formulation (2.1.102). l_i^k are the basis polynomials composing Π_k.

$$x_{n+1-i} \quad \text{with} \quad i = 1, \ldots, k$$

$$\Pi_k(t) = \sum_{i=1}^{k} l_i^k(t)x_{n+1-i} \tag{2.1.102}$$

The derivatives of x_{n+1-i} are expressed by (2.1.103).

$$\dot{x}_{n+1-i} = f(t_{n+1-i}, x_{n+1-i}) \tag{2.1.103}$$

By extrapolation of this polynomial and introducing the coefficients α_{k-i}^P, the next solution point can be predicted:

$$\alpha_{k-i}^P = -l_i^k(t_{n+1}) \tag{2.1.104}$$

$$x_{n+1}^P = \Pi_k(t_{n+1}) = -\sum_{i=1}^{k} \alpha_{k-1}^P x_{n+1-i} \tag{2.1.105}$$

Subsequently, this prediction is used in a predictor-corrector-scheme following the form in (2.1.108). h_n is the step size of the solver steps. For the corrector step the coefficient α_{k-i}^C and the coefficient β_k^C are determined by:

$$\alpha_{k-i}^c = \frac{\dot{l_i^{k+1}}(t_{n+1})}{\dot{l_0^{k+1}}(t_{n+1})} \tag{2.1.106}$$

$$\beta_k^c = \frac{1}{h_n \dot{l_0^{k+1}}(t_{n+1})} \tag{2.1.107}$$

The corrector step can be performed using Newton's method. Therefore, the BDF is suitable for stiff problems. This makes the method suitable for solving MBS systems.

$$\begin{aligned}
\text{Predict:} \quad & x_{n+1}^P = -\sum_{i=1}^{k} \alpha_{k-1}^P x_{n+1-i} \\
\text{Evaluate:} \quad & \dot{x}_{n+1} = f(t_{n+1}, x_{n+1}^P) \\
\text{Correct:} \quad & x_{n+1} = x_{n+1}^P - \sum_{i=1}^{k} \alpha_{k-1}^C x_{n+1-i} + h_n \beta_k^C \dot{x}_{n+1} \\
\text{Evaluate:} \quad & \dot{x}_{n+1} = f(t_{n+1}, x_{n+1})
\end{aligned} \tag{2.1.108}$$

Differential/Algebraic System Solver

SIMPACK's default solver for time integration is SODASRT2 which is based on the solver algorithm DASSL. The method of DASSL is explained by [11, 32]. This solver is applicable even if the system of equations can not be written in the explicit ODE form $x' = f(t, x)$ or if it was at least impractical to do so. In general, a system of equations can be expressed as (see also Sect. 2.1.1):

$$F(t, x, x') = 0 \tag{2.1.109}$$

$$x(t_0) = x_0 \tag{2.1.110}$$

$$x'(t_0) = x'_0 \tag{2.1.111}$$

The basic principle is to replace the derivative x' by an approximation using BDF, whereas the order of the approximating polynomial and the step size is variable in each iteration. That way the non-linear equation can be rewritten as:

$$F(t, x, \hat{\alpha}x + \beta) = 0 \tag{2.1.112}$$

As explained in abstract regarding BDF Newton's method in a modified version solves this corrector equation.

$$x^{k+1} = x^k - c(\frac{\partial F}{\partial x'} + \alpha\frac{\partial F}{\partial x})^{-1} F(t, x^k, \hat{\alpha}x^k + \beta) \tag{2.1.113}$$

The iteration matrix defined by the term $\boldsymbol{G} = c(\frac{\partial F}{\partial x'} + \hat{\alpha}\frac{\partial F}{\partial x})^{-1}$ in (2.1.113) is used for as many time steps as possible. The factor c is chosen to speed up the calculation and depends on the ratio of the current coefficient $\hat{\alpha}$ and the coefficient α from the iteration step, where the matrix $\boldsymbol{G}$ was calculated last. However, the solver might not converge if $\hat{\alpha}$ and α differ too much.

Discontinuity Handling

As shown before, the equations of motion are solved by numerical time integration which requires continuity also in their derivatives. However, the equations of motion may contain discontinuities due to impact phenomena that lead to jump discontinuities in velocities, dry friction, or discontinuous time-dependent input functions [11]. These discontinuities lead either to a solver failure due to step size control failure or wrong error estimation or at least inefficient solving.

In many cases, the equations of motion are piecewise continuous. Thus, they can be defined in dependency on a switching function θ. The roots of this switching function describe whether discontinuities appear in dependency of states and time.

$$\theta(t, x(t)) = (\theta_1(t, x(t)), ..., \theta_n(t, x(t)))^T \tag{2.1.114}$$

The system's equations can therefore be written as:

$$\dot{x} = f(t, x, \text{sign}\theta) \tag{2.1.115}$$

It can be seen that the equations depend on the sign of the components of $\theta(t, x(t))$. Thus, the solution for a general differential equation with a discontinuity occurring at time $\hat{t}$ is composed of two branches [11]:

$$x(t) = \begin{cases} x^-(t) \; t \le \hat{t} \\ x^+(t) \; t > \hat{t} \end{cases} \tag{2.1.116}$$

If the system's equations contain switching functions, the signs of the components of $\theta(t, x(t))$ are checked for a change in sign after each integration step. If no sign changes, the integration is continued, otherwise the discontinuity is localised and the variables of the right hand side of the equations of motions are adjusted according to the root states of the switching functions. The integration is restarted afterwards.

The present section summed up the mathematical description of mechanical systems as basis of the MBS, the implementation of flexible bodies to a mechanical system and the solving of such mechanical system equations. This builds the theoretical basis of the valve dynamic analysis.

The following section shows statistical methods. The statistical evaluation of data and the consideration of statistical variations of boundary conditions and load data increase the reliability of the fatigue strength assessment.

2.2 Statistical Evaluation

The application of statistical methods either to evaluate measurement data or to increase information obtained from simulations is elementary when it comes to strength assessment. Statistical fluctuations have an impact on both material data and load data.

This section presents the main statistical methods used in this thesis. In this context, the first part focuses on the measurement data evaluation. Therefore, the assessment of the measurement data distribution and the estimation of the data maximum is a main part in the subsection.

Furthermore, methods of sensitivity analysis are focus of the second part of this section.

2.2.1 Statistic Methods for Measurement Data Evaluation

This subsection introduces the normal distribution and the extreme value distribution for the statistical evaluation of measurement data. Furthermore, the determination of confidence intervals for unknown distributions is part of this subsection.

2.2.1.1 Distributions

The determination of the distribution of measurement data enables the evaluation of important data for the fatigue strength assessment.

The distribution widely known and often used is the Gaussian normal distribution. This distribution applies to several random variables in various applications. The central limit therorem (CLT) provides the reason for that. According to [19], the law of large numbers claims that the mean value $\bar{X}$ of independent and identically distributed random variables $X_1, ...X_n$ concentrates around the common expected value μ_M with increasing number n_M of random variables. Beyond that, the CLT states the distribution shape of the standardized random variables. The standardised random variable Z_n has an expected value of 0 and a standard deviation σ_M of 1. It is therefore defined by:

$$Z_n = \sqrt{n} \cdot \frac{\bar{X}_n - \mu}{\sigma} \tag{2.2.1}$$

The CLT declares that the distribution follows the standard normal distribution.

This means that the sum of identical distributed random variables is approximately normally distributed [26]. For this reason, the normal distribution of random variables is often assumed in various use cases.

Nevertheless, if the distribution regions of interest are maximum or minimum values of a sample, the normal distribution may not be a sufficient approximation. Especially at the upper and lower tails of the normal distribution, the real distribution of random variables can differ significantly from the normal distribution.

The field of extreme-value theory focuses on the statistical description of these extremal events. The following explanations on extreme value theory are mainly based on [12].

For the description of a random variable distribution beyond a specific probability limit, the first step is to determine the probability of the excess of that threshold. The mean excess function F_u, given by (2.2.2), describes the distribution of the random variable X over a threshold u_{tr} by the probability P.

$$F_u(x) = P(X - u_{tr} \leq x | X > u_{tr}), \qquad x \geq 0 \tag{2.2.2}$$

The distribution of values exceeding a high threshold can be described by the generalised Pareto distribution (GPD) G_ξ [12]. The GPD is defined by (2.2.3) with the form parameter ξ.

$$G_\xi(x) = \begin{cases} 1 - (1 + \xi x)^{-\frac{1}{\xi}}, & \text{if} \quad \xi \neq 0, \\ 1 - e^{-x}, & \text{if} \quad \xi = 0 \end{cases} \tag{2.2.3}$$

The equation (2.2.5) resulting from (2.2.2) and (2.2.3) describes the distribution beyond a threshold in good approximation [31]. Therefore, an estimation of the deviation parameter $\tilde{\sigma}$ of the distribution over the threshold is utilized. N_{tot} is the total number of samples, while N_u is the number of samples greater than the threshold u_{tr}. $\tilde{\mu}$ is the ratio of samples over the threshold and the total number of samples:

$$\tilde{\mu} = \frac{N_u}{N_{tot}} \tag{2.2.4}$$

With $\tilde{\mu}$ the function describing the extreme-value distribution above the threshold holds [31]:

$$F(x) \approx 1 - \tilde{\mu} \cdot \left(1 - \frac{\xi(x - u_{tr})}{\tilde{\sigma}}\right)^{\frac{1}{\xi}}, \tag{2.2.5}$$

$$x \geq u_{tr} \quad \text{for} \quad \xi < 0,$$

$$u_{tr} \leq x \leq \frac{\tilde{\sigma}}{\xi} + u_{tr} \quad \text{for} \quad \xi > 0,$$

The threshold u_{tr} depends on the specific data. In general, the threshold must be high enough or low enough respectively to ensure that the GPD describes the distribution properly, on the one hand. On the other hand, there still needs to be enough data to estimate the distribution sufficiently.

The parameters ξ and $\tilde{\sigma}$ can be estimated by the method of moments (MOM). This method expresses the parameter to estimate by the moments of the distribution. The term moment in the context of stochastic describes a measure to characterise a distribution. The first moment is the expectation value of a distribution. Higher order moments are the variance, the skewness, and kurtosis. These theoretical moments are then replaced by the empirical moments. The empirical moments are directly derived from the sample. An example is the expectation value that is replaced by the mean value of the sample.

The method of probability-weighted moments provides another estimation of parameters. For the application on valve load data, this method does not deliver sufficient results. [31] shows that the method of probability-weighted moments underestimates the maximum valve load for certain engine operation points. The MOM never underestimates the valve load for the evaluated data and is therefore more on the safe side regarding load data determination. References [12, 19] provide further information on parameter estimation.

2.2.2 *Design of Experiments*

The output of a system often depends on several influencing factors. In this context, some influencing factors have a higher impact on the output value, some influencing factors have a minor influence on the output value. With a targeted variation of influencing factors and evaluation of the changes in output, the specific influence of each factor can be determined. The methods of DoE provide various strategies on the determination of factor influence. The influence of a factor can be linear, quadratic or even higher order. A linear influence can be determined by evaluating the output for two levels of the influencing factor, for example the defined maximum

Table 2.1 Factor influence on output value

Influencing factor		Output value
Load	Clearance	Max. Strain
0	0	1.0
+	0	1.5
0	+	2.0
+	+	2.2

and minimum of the factor. A quadratic influence requires the evaluation of at least three different values which the influencing factor can take. These are the factor levels. Moreover influencing factors can interact with each other regarding their effect on the output value. If several factors influence an output value, the number of necessary evaluations increases significantly. Therefore, there are strategies that provide almost the same information on the factor influences but omit some of the evaluations.

This section introduces the strategies that are suitable for the factor influence analysis of the mechanical valve train system.

The output of interest in this context is the maximum valve load. The load is determined by the valve strain as this can be determined by simulation and measurement. The maximum load of the valve depends on several factors, e.g. the engine load or the valve clearance. This is true for the real component as well as for a simulation model. For many use cases the factors' magnitude of influence on an output value is unknown.

In the context of the possible influences on the valve load, Table 2.1 shows an example on interaction of influencing factors: The influencing factors are the previously introduced engine load and the valve clearance. The output value is the maximum valve strain. If each of the influencing factors is changed to its maximum value (+), the valve strain increases. The influence of the factor valve clearance is higher, as the output value is higher for the maximum clearance than for the maximum load. When both factors are changed to their maximum together, the strain value increases but not linearly, compared to the strain value with only one factor at its maximum. This example shows the impact of interactions, as the influences of single factors can not be simply added up to determine the overall influence of all factors.

There are two main kinds of experimental designs. On the one hand, there are screening designs enabling the evaluation of a high number of influencing factors. On the other hand, there are detailed designs enabling the evaluation of non-linear correlations of factors and output values. The following subsection describes the experimental designs that are important in this work.

The terms *main effect* and *interaction effect* describe the kind of factor influence on an output attribute Y. Main effects are the effects of a single factor on the attribute, while the interaction effects quantify the influence of a combination of factors. With the method of DoE, the relation between factors and model attributes is described by a mathematical model in terms of a polynomial. Thus, a linear model is defined by (2.2.6), [43]. n_f is the number of factors, the coefficients c_i are polynomial coefficients of the main effects, c_{ij} are polynomial coefficients of the interaction effects, c_0 is the constant of the polynomial. ϵ is the error of the model compared to the real relation between output and factors. Ideally, ϵ is zero. x_f is the variable of the polynomial and therefore describes all the possible factor values of the model.

$$Y = c_0 + \sum_{i=1}^{n_f} c_i x_{f,i} + \sum_{i=1}^{n_f-1} \sum_{j=i+1}^{n_f} c_{ij} x_{f,i} x_{f,j} + \epsilon \tag{2.2.6}$$

Here, the first term on the right side is the constant term of the polynomial, the second term describes the main effects and the third term describes the interaction effects on the model attribute Y.

With increasing polynomial order, terms describing higher order main effects are added up to the linear model, resulting in (2.2.7). The general equation of DoE models can be written as:

$$Y = c_0 + \sum_{i=1}^{n_f} c_{1,i} x_{f,i} + \sum_{i=1}^{n_f} c_{2,i} x_{f,i}^2 + \dots + \sum_{i=1}^{n_f} c_{k,i} x_{f,i}^k + \sum_{i=1}^{n_f-1} \sum_{j=i+1}^{n_f} c_{w,ij} x_{f,i} x_{f,j} + \epsilon \tag{2.2.7}$$

Higher order interactions do not affect the model attribute significantly for most cases and are therefore not considered [43].

The classic experimental design is the full-factorial design. It includes all combinations of all factors on all levels. Therefore, this design delivers the most information but needs many experiments. Especially for higher order and interaction influences, the number of necessary experiments increases rapidly.

Thus, there are experimental designs more suitable for the evaluation of a quadratic or even a higher order model because they need fewer experiments for the model generation and provide almost as much information as the full factorial design. One of the designs suitable for a quadratic model is the central composite design (CCD). The CCD consists of a two-level plan, also referred to as cube in graphical representations. This plan is complemented by centre points at the medians of all factors. Furthermore, axial points are added. Based on the centre point, one factor is varied above and below the levels of the two-level plan. Therefore, each factor is varied on five levels. The graphical representation for a three-factor CCD is shown in Fig. 2.5. Here, the violet mid-point represents the experiment with each factor on its mid level. The green points represent the cube of the experimental design, where each factor only has two levels. There are eight points because the variation of three factors on two levels leads to overall $2^3 = 8$ possible combinations. The orange points are on higher and lower levels than the cubic design points and represent the so-called star. These higher

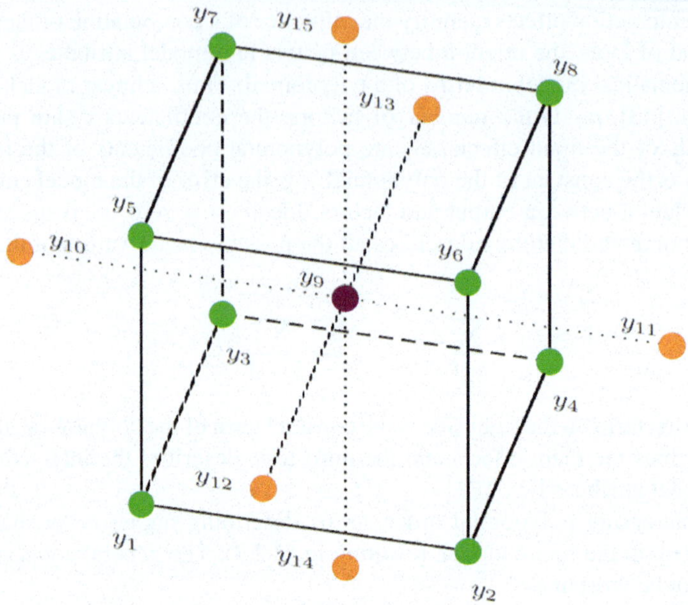

Fig. 2.5 Central composite design with three factors, [43]

level points are only analysed regarding their main effects. Therefore, there are six points for three levels. Another design for higher order models enabling a reduction of experiments is the latin hypercube design (LHD) based sampling. For the LHD based sampling. The value range of each factor is divided by the sample size. For a design with 100 samples this leads to 100 factor levels. Each factor level appears only once in the whole design. The factor levels are permuted in the design matrix, avoiding that one factor on its ith level is combined with another factor on its ith level. This way, the design space is filled and factor correlations are nearly excluded from the experimental design. This approach appears to be up to 50% more efficient than the Monte-Carlo sampling method that is based on random sampling [43, 44].

The number of necessary experiments is strongly dependent on the number of factors to evaluate. For a large number of factors it may be sufficient to eliminate some of the less important factors by applying a screening design. Screening designs have a smaller number of combinations taken for experiments from all possible combinations. Therefore, they do not deliver quantitative results of factor influences. However, screening designs deliver qualitative results with only a small number of samples and thus enable the elimination of less important factors.

A method to classify factors is the Morris method described in [27]. The classes of factors in this method are factors with negligible effects on the model attribute, factors with linear effects on the model attribute and factors with non-linear effects on the model attribute. This method is part of the field of one factor at a time (OAT) designs.

The Morris method provides an effective way to determine the most important factors by running only a small number of experiments [5].

Firstly, the overall number of OAT designs must be defined. Each factor is discretised depending on the previously defined number of experiments. Then, a start combination of all factors is randomly determined as a first experiment. After this, for every following experiment one factor is changed randomly in the given input space until all factors are changed once with respect to the start combination. Therefore, each OAT design includes $n_f + 1$ experiments, where n_f is the number of factors. By repeating this OAT elementary effects E_j can be estimated by [21]:

$$E_j^{(i)} = \frac{f(X_f^{(i)} + \Delta_s e_j) - f(X_f^{(i)})}{\Delta_s} \qquad j = 1, .., n_f \tag{2.2.8}$$

X_f is the vector of inputs, Δ_s is the increment size defined as a multiple of the term $\frac{1}{n_l - 1}$ with the number of factor level n_l. e_j is a vector of the canonical base.

The evaluation of parameter influences uses the mean μ_j^* and the standard deviation σ_j of the elementary effects $E_j^{(i)}$. Parameters with a higher mean of elementary effects have more influence on the output value than parameters with a smaller mean of elementary effects. The ratio of standard deviation and mean of elementary effects of a parameter characterises the kind of influence the parameter has on the output value. Small ratios show a constant or linear influence, higher ratios of standard deviation and mean of elementary effects show a non-linear influence or high interaction effects with other parameters.

2.3 Fatigue Strength Analysis

This thesis follows the well established FKM guideline regarding fatigue strength assessment. This section gives an overview over the procedure in the guideline. The FKM guideline provides two methods of fatigue strength assessment of dynamically loaded components. The present work focuses on the dynamic fatigue strength assessment of steel components with local stresses. A linear-elastic material behaviour is presumed by the guideline. The following description of the fatigue strength assessment is based on [18, 34].

Firstly the guideline categorises occurring stresses into three classes:

Proportional Stresses Proportional stresses lead to constant principal stress directions and often result from a single load acting on the component. If the load varies over time, the multi axial stresses are proportional to each other.

Synchronous Stresses Synchronous stresses have constant stress amplitudes with varying mean stresses and can result from the combination of a constant load with a time-varying load.

Non-proportional Stresses Two or more time-varying loads can result in non-proportional stresses with principal stresses varying in direction and amplitude.

The fatigue strength assessment compares the local stress values with the values of the component fatigue strength under consideration of complying safety factors.

The component fatigue strength assessment sets off material properties against design parameters to evaluate component specific strength values.

On the load side, the load collective describing the frequency distribution of load cycles needs to be determined. Usually the rainflow counting method is applied to determine the load collective from load-time signals. The special case of a one-step collective occurs for a constant amplitude. For steel, the guideline assumes a component S-N curve with a knee point cycle number N_D, so that the required cycle number for fatigue limit safety is $N \geq N_D$.

If the load amplitude varies, the load collective has multiple levels. The complex process of converting the loads to a common load ratio R_σ and sorting the load collective steps by size can be simplified by using the damage-equivalent stress amplitude. By using this damage-equivalent amplitude, the fatigue strength assessment becomes the same type like the fatigue limit assessment and has the same quality like the above described fatigue strength assessment process.

The damage-equivalent stress amplitude σ_{equi} equals the damage of the load collective and the corresponding knee point of the component's constant S-N curve is equal to the assigned cycle number $\bar{N}$. The damage-equivalent stress amplitude is defined as the quotient of the load collective maximum $\sigma_{a,1}$ and the amplitude strength factor K_{BK}:

$$\sigma_{equi} = \frac{\sigma_{a,1}}{K_{BK}} \tag{2.3.1}$$

At this, the amplitude strength factor K_{BK} depends on the assigned load cycle, the knee point and the slope k of the component constant S-N curve.

$$K_{BK} = \left(\frac{N_D}{\bar{N}}\right)^{\frac{1}{k}} \tag{2.3.2}$$

The determination of the damage-equivalent amplitude based on the consistent version of Miner's rule is iterative by approximating the assigned cycles $\bar{N}$.

The resulting damage-equivalent amplitude is used for the calculation of the endurance safety factor in the fatigue strength assessment equivalent to the fatigue limit assessment. In contrast to the actual operational loads, the variance of the damage-equivalent amplitude is small and can be determined within a short measurement period.

On the component strength side, the assessment requires the determination of the material fatigue limits. Unless specific data for the material fatigue limit can directly be determined from material tests, the material fatigue limits can be derived from the material tensile strength R_m, see (2.3.3). Therefore, the fatigue strength factor depending on the material class for completely reversed normal stress $f_{W,\sigma}$ is provided by tabular data within the guideline. With this factor the stress limit of completely reversed normal stress $\sigma_{W,zd}$ yields:

$$\sigma_{W,zd} = f_{W,\sigma} \cdot R_m \tag{2.3.3}$$

Following the von Mises criterion, the completely reversed shear fatigue limit $\tau_{W,s}$ is lower than the axial fatigue limit by a factor of $f_{W,\tau} = \sqrt{3}$. Thus, the shear fatigue limit is defined by (2.3.4).

$$\tau_{W,s} = f_{W,\tau} \cdot \sigma_{W,zd} \tag{2.3.4}$$

As a next step, the design factors $K_{WK,\sigma}$ and $K_{WK,\tau}$ are evaluated by determining and considering the notch geometry at the critical point, the surface roughness, the surface treatment and the surface coating. Subsequently, these design factors and the previously evaluated material fatigue limits lead to the component specific fatigue limits for normal stress σ_{WK} and shear stress τ_{WK}:

$$\sigma_{WK} = \frac{\sigma_{W,zd}}{K_{WK,\sigma}} \tag{2.3.5}$$

$$\tau_{WK} = \frac{\tau_{W,s}}{K_{WK,\tau}} \tag{2.3.6}$$

The next step determines the mean stress sensitivity that describes how much the mean stress affects the component fatigue limit. The required factors a_M and b_M are fixed values for steel and provided by the FKM guideline. The mean stress sensitivity of normal stress M_σ and shear stress M_τ is defined by:

$$M_\sigma = a_M \cdot 10^{-3} \cdot \frac{R_m}{\text{MPa} + b_M} \tag{2.3.7}$$

$$M_\tau = f_{W,\tau} \cdot M_\sigma \tag{2.3.8}$$

Furthermore, the stress relation R_σ of upper stress boundary to lower stress boundary defines the field of mean stress as shown in Table 2.2. The visualization of the mean stress fields in the Haigh diagram is shown in Fig. 2.6.

The guideline distinguishes the following four cases of overload for the determination of the mean stress influence factor. σ_m is the mean stress, σ_{min} is the minimum stress, σ_{max} is the maximum stress.

- $\sigma_m = \text{const.}$
- $R_\sigma = \text{const.}$
- $\sigma_{min} = \text{const.}$
- $\sigma_{max} = \text{const.}$

Table 2.2 Fields of mean stress

Stress ratio R_σ	Field of mean stress
$R_\sigma = \pm\infty$	Zero compression stress
$R_\sigma > 1$	Fluctuating compression stress
$R_\sigma = -1$	Completely reversed stress
$R_\sigma > -1$	Alternating compression stress
$R_\sigma = 0$	Zero tension stress
$R_\sigma \geq 0,5$	High fluctuating tension stress

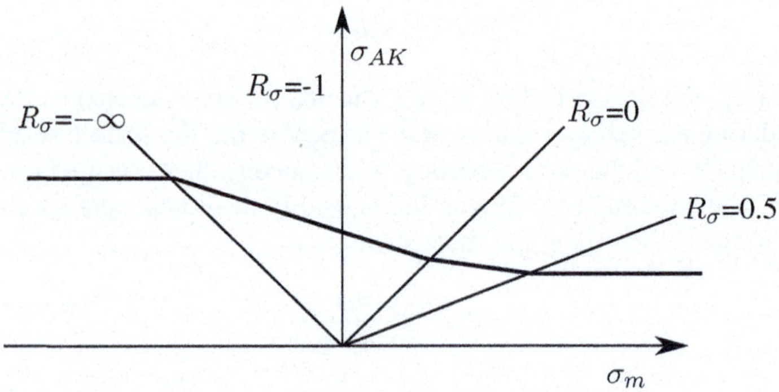

Fig. 2.6 Haigh diagram for normal stresses

The mean stress influence factor K_{AK} is then determined considering the overload case, the mean stress sensitivity and the field of mean stress.

Furthermore, the guideline determines the safety factor. It results from the load safety factor j_S and the material safety factor j_F which is determined by the regularity of component inspections and the severity of consequences of component failure. Furthermore, the temperature factor $K_{T,D}$ defined by (2.3.9) influences the fatigue safety factor j_D.

$$K_{T,D} = 1 - 10^{-3} \cdot \frac{T}{C^\circ} \tag{2.3.9}$$

$$j_D = j_S \frac{j_F}{K_{T,D}} \tag{2.3.10}$$

The assessment is finished by calculating the cyclic degree of utilisation a_{BK} which is defined as the relation of the stress amplitude and the amplitude fatigue strength of the component. The latter takes the safety factor into consideration. To meet the conditions of endurance strength, the cyclic degree of utilisation should be equal or less than one. The software FEMFat calculates the safety factor of fatigue strength as the result of the fatigue strength assessment which is the reciprocal of the cyclic degree of utilisation in this context. As a consequence, the safety factor has to be greater than one. However, usually higher safety factors are required by company internal guidelines.

$$a_{BK} = \frac{\sigma_{equi}}{(K_{BK} K_{AK} \sigma_{WK}) j_D} \tag{2.3.11}$$

Another important criteria is the failure probability of components. In literature, there are different suggestions on a sufficient failure probability. MAN ES aims for a component failure probability of $P_{A,log} \leq 10^{-6}$, meaning that less than one part per one million parts fails.

As the damage-equivalent amplitude is assumed to cover all appearing loads during the required running time of the component, the standard deviation of the load amplitude is assumed to be zero. This assumption is in accordance with the FKM guideline. According to [18], the failure probability is defined as:

$$P_A = \frac{1}{\sqrt{2 \cdot \pi}} \cdot \int_{-\infty}^{u_0} e^{-\frac{u^2}{2}} du \tag{2.3.12}$$

The failure probability is therefore a function of u_0 that is defined by the mean m_F and the standard deviation s_F of the material's distribution.

$$u_0 = -\frac{m_F}{s_F} \tag{2.3.13}$$

For the logarithmic normal distribution the safety factor j_{FS} can also be expressed as:

$$j_{FS} = 10^{-u_0 \cdot s_F} \tag{2.3.14}$$

Because u_0 is directly related to the failure probability $P_{A,log}$, this equation relates the failure probability to the safety factor.

References

1. ARNOLD, M. : *Simulation algorithms in vehicle system dynamics*. Univ., Fachbereich Mathematik und Informatik, 2004
2. ARNOLD, M. ; CARRARINI, A. ; HECKMANN, A. ; HIPPMANN, G. : Simulation Techniques for Multidisciplinary Problems in Vehicle System Dynamics. In: *Vehicle System Dynamics Supplement 40* (2004), 17–36
3. BARTEL, D. : *Simulation von Tribosystemen: Grundlagen und Anwendungen*. 1. Wiesbaden : Vieweg+Teubner, 2010
4. BRENAN, K. E. ; CAMPBELL, S. L. ; PETZOLD, L. R.: *Numerical solution of initial value problems in differential algebraic equations*. New York, NY : North Holland, 1989
5. CAMPOLONGO, F. ; CARIBONI, J. et al.: Sensitivity analysis: how to detect important factors in large models. In: *European Commission, Joint Research Centre, Ispra (VA), Italy* (2007)
6. CRAIG, R. R. ; BAMPTON, M. C. C.: Coupling of substructures for dynamic analyses. In: *AIAA Journal* 6 (1968), Nr. 7, 1313–1319
7. DASSAULT SYSTÈMES SIMULIA CORP.: *Simpack - SIMULIA User Assistance 2022x*. 2021
8. DIETZ, S. ; WALLRAPP, O. ; WIEDEMANN, S. : Nodal vs. Modal Representation in Flexible Multibody System Dynamics. In: JORGE A. C. AMBRÓSIO (ed.): *Multibody Dynamics 2003*, 2003
9. EBERHARD, P. ; SCHIEHLEN, W. : Computational Dynamics of Multibody Systems: History, Formalisms, and Applications. In: *Journal of Computational and Nonlinear Dynamics* 1 (2005), Nr. 1, 3–12
10. EBRAT, O. ; MOURELATOS, Z. P. ; VLAHOPOULOS, N. ; VAIDYANATHAN, K. : Calculation of Journal Bearing Dynamic Characteristics Including Journal Misalignment and Bearing Structural Deformation. 47 (2004), Nr. 1, 94–102
11. EICH- SOELLNER, E. ; FÜHRER, C. : *Numerical Methods in Multibody Dynamics*. Stuttgart : Teubner, 1998
12. EMBRECHTS, P. ; KLÜPPELBERG, C. ; MIKOSCH, T. : *Modelling Extremal Events*. Berlin, Heidelberg : Springer Berlin Heidelberg, 1997
13. FISCHER, DR. P. ; WITTEVEEN, W. ; SCHABASSER, M. : Integrated MBS – FE – Durability Analysis of Truck Frame. In: *Proceedings of the 15th European ADAMS Users Conference, Rome* (2000)
14. GASCH, R. ; KNOTHE, K. ; LIEBICH, R. : *Strukturdynamik*. Berlin, Heidelberg : Springer Berlin Heidelberg, 2012
15. GREENWOOD, J. A. ; TRIPP, J. H.: The Contact of Two Nominally Flat Rough Surfaces. In: *Proceedings of the Institution of Mechanical Engineers* 185 (1970), Nr. 1, 625–633.
16. GRUBER, K. ; RUDER, H. ; DENOTH, J. ; SCHNEIDER, K. : A comparative study of impact dynamics: wobbling mass model versus rigid body models. In: *Journal of Biomechanics* (1998), Nr. 31, 439–444
17. GUYAN, R. J.: Reduction of stiffness and mass matrices. In: *AIAA Journal* 3 (1965), Nr. 2, 380.
18. HAIBACH, E. : *Betriebsfestigkeit: Verfahren und Daten zur Bauteilberechnung*. 3. Berlin : Springer, 2006 (VDI-Buch)
19. HARTUNG, J. : *Statistik*. Oldenbourg Wissenschaftsverlag, 2009
20. HIPPMANN, G. : An algorithm for compliant contact between complexly shaped surfaces in multibody dynamics. In: JORGE A. C. AMBRÓSIO (ed.): *Multibody Dynamics 2003*, 2003
21. IOOSS, B. ; LEMAÎTRE, P. : A review on global sensitivity analysis methods. In: DELLINO, G. (ed.) ; MELONI, C. (ed.): *Uncertainty Management in Simulation-Optimization of Complex Systems*. Springer US, 2015
22. KNOLL, G. ; BRANDS, C. : Forschungsvorhaben 614: Rechnergestützte Simulation der Dynamik und der Beanspruchung des Ventiltriebs. In: *FVV Forschungsheft* 661 (1998)
23. KÖHLER, E. : *Verbrennungsmotoren*. Wiesbaden : Springer Fachmedien, 2009 (ATZ / MTZ-Fachbuch)
24. KOUTSOVASILIS, P. : *Model Order Reduction in Structural Mechanics Coupling the Rigid and Elastic Multi Body Dynamics*. Dresden, Technische Universität Dresden, Dissertation, 2009

25. LANG, O. R. ; STEINHILPER, W. : *Gleitlager: Berechnung und Konstruktion von Gleitlagern mit konstanter und zeitlich veränderlicher Belastung*. Bd. 31. Berlin Heidelberg : Springer-Verlag, 1978
26. LANGE, T. ; MOSLER, K. : *Statistik kompakt*. Berlin, Heidelberg : Springer Berlin Heidelberg, 2017
27. MORRIS, M. D.: Factorial Sampling Plans for Preliminary Computational Experiments. In: *Technometrics* 33 (1991), Nr. 2, 161–174
28. MÜLLER, W. H. ; FERBER, F. : *Technische Mechanik für Ingenieure*. Fachbuchverlag Leipzig im Carl Hanser Verlag, 2011
29. OFFNER, G. ; HASLINGER, J. ; SOPOUCH, M. ; ZINKIEWICZ, B. : Linearized Modal Analysis of Vehicle Powertrains. In: *8th ECCOMAS Thematic Conference on Multibody Dynamics*, 2017
30. PARIKYAN, T. : Complex modes analysis for powertrain and driveline applications. In: *Proceedings of International Conference on Noise and Vibration Engineering (ISMA)* (2016), 2761–2774
31. PAUSIN, A. ; BECK, A. ; BÖHM, P. ; SPECKERT, M. ; FIEDLER, J. : Ventiltriebsmessungen von Großmotoren zur Lastermittlung. In: *DVM-Workshop Zuverlässigkeit und Probabilistik* (2021)
32. PETZOLD, L. R.: A description of DASSL: A differential/algebraic system solver, North-Holland Publishing Company
33. POPOV, V. L.: *Kontaktmechanik und Reibung*. 3. Wiesbaden, Germany : Springer, 2016
34. RENNERT, R. ; KULLIG, E. ; VORMWALD, M. ; ESDERTS, A. ; SIEGELE, D. : *Rechnerischer Festigkeitsnachweis für Maschinenbauteile: Aus Stahl, Eisenguss- und Aluminiumwerkstoffen*. 6. Frankfurt (Main) : VDMA, 2012
35. RIEß, A. : *Model order reduction based simulation and optimization of large bore internal combustion engines*. Universität Augsburg, Mathematisch-Naturwissenschaftlichen Fakultät, Dissertation, 2014
36. ROBERSON, R. E. ; SCHWERTASSEK, R. : *Dynamics of Multibody Systems*. Berlin, Heidelberg : Springer Berlin Heidelberg, 1988
37. SCHIEHLEN, W. : Multibody System Dynamics: Roots and Perspectives. In: *Multibody System Dynamics* (1997), Nr. 1, 149–188
38. SCHIEHLEN, W. : Computational dynamics: Theory and Applications of Multibody Systems. In: *European Journal of Mechanics - A/Solids* 25 (2006), Nr. 4, 566–594
39. SCHWERTASSEK, R. ; WALLRAPP, O. : *Dynamik flexibler Mehrkörpersysteme*. Wiesbaden : Vieweg+Teubner Verlag, 1999
40. SCHWERTASSEK, R. ; DOMBROWSKI, S. V. ; WALLRAPP, O. : Modal Representation of Stress in Flexible Multibody Simulation. In: *Nonlinear Dynamics* 20 (1999), Dez., Nr. 4, 381–399
41. SCHWERTASSEK, R. ; WALLRAPP, O. ; SHABANA, A. A.: Flexible Multibody Simulation and Choice of Shape Functions. In: *Nonlinear Dynamics* 20, Nr. 4, 361–380
42. SHABANA, A. ; SCHWERTASSEK, R. : Equivalence of the floating frame of reference approach and finite element formulations. In: *International Journal of Non-Linear Mechanics* 33 (1998), Nr. 3, 417-432.
43. SIEBERTZ, K. ; BEBBER, D. ; van HOCHKIRCHEN, T. : *Statistische Versuchsplanung*. Berlin, Heidelberg : Springer Berlin Heidelberg, 2017
44. STOCKINGER, A. ; WARTZACK, S. : Application of Sampling Methods for Statistical Tolerance Analysis. In: *7. Weimarer Optimierungs- und Stochastiktage*, 2010
45. WALLRAPP, O. ; WIEDEMANN, S. : Comparison of Results in Flexible Multibody Dynamics Using Various Approaches. In: *Nonlinear Dynamics* 34, Nr. 1, 189–206
46. ZACHMANN, G. : Rapid Collision Detection by Dynamically Aligned DOP-Trees. In: IEEE (ed.): *Virtual Reality Annual International Symposium*, 1998

Chapter 3
State of the Art

This chapter contextualizes the aim of this thesis, the assessment of valve fatigue strength, to the current state of the art. The aim of this thesis holds two different topics. One topic is the determination of valve fatigue strength, the other topic is the simulative determination of the loads that go to the fatigue strength assessment. To contextualize this thesis to the current state of the art, these two topics are considered separately. Moreover, the selected approach of the thesis, the predictive simulative approach is also considered in the contextualization of the thesis.

The chapter is divided into three sections. The first section focuses on the simulation of valve train systems. The second section focuses on the fatigue strength assessment of the valves. This does not necessarily go along with a simulative approach. The last section introduces the measurement approaches on valve dynamic. This section also describes the current fatigue assessment at MAN ES, which is mainly based on measurements.

3.1 Simulation of Valve Train and Valve Dynamics

The valve train dynamic in general has been researched in several publications. The focus of these publications varies from valve stress determination to valve wear analysis and the analysis of valve train bearing loads. There is a strong focus on passenger car engines, because the majority of load exchange valves are installed in these engines.

With respect to simulation method, the main focus is on MBS and FEM modelling. The method of MBS is explained in Sect. 2.1. The FEM approach is also a numerical procedure solving differential equations. It uses a subdivision of structures into simple shapes (hexaeder or tetraeder for 3D geometries) and therefore discretizes the continuous structure. Because of the simple geometry of the elements, their behaviour can be assessed by simple initial functions. [2, 18] give a detailed overview over the FEM approach. The main difference between the MBS and the FEM approach is

A. Eret, *Dynamic and Fatigue Assessment of Heavy-Duty Engine Valves*, Mechanics and Adaptronics, https://doi.org/10.1007/978-3-031-49151-1_3

that FEM is mainly used for static problems or dynamic problems in the frequency domain. It is suitable to simulate specific events, e.g. the valve closing. It is not suitable for big deformations, that occur during the valve working cycle while MBS is used for dynamic problems in the time domain when big deformations of the system appear.

The valve body, the valve seat ring, valve springs and the valve guidance are components in direct contact with the valve and therefore influence the valve dynamic. The modelling depth of these components influences the accuracy of the simulation significantly. Depending on the specific aim, valve train simulation models vary from rigid models with simple spring-damper contact to full flexible models with elastic contact formalism.

The investigations in [19] pursue a target similar to the valve dynamic analysis aim of the present thesis. The research project analyses the valve dynamic and valve stresses for an accurate calculation of a reliable valve design and prevention of valve damage. This publication provides the overall highest modelling depth regarding the MBS model in combination with measurement based validation. The model includes the valve, the valve spring, the valve guidance and the valve seat ring. These components are outlined in following subsections. Furthermore, the model includes the whole actuation mechanics of the valve train, starting from the overhead camshaft and continuing to the valve. The roller cam follower is a rigid body in this implementation. The camshaft is a reduced beam structure showing the bending and torsional vibration of the camshaft.

The MBS model also includes flexible bodies. These are the valve, valve seat ring and guidance body. The contact between valve and guidance is implemented with an EHD formulation to consider interactions of lubrication film and solid states. The spring is implemented as an uniaxial series connection of spring-damper elements. The modelling depth of components in direct contact with the valve is high due to the proclaimed target of valve stress determination for the purpose of a reliable valve design. The validation uses experimental findings from measurements on a towed and unfired engine.

Another detailed investigation of valve dynamic is published by [26]. The thesis describes the valve train of a passenger car with an overhead camshaft. The simulation model therefore includes the same components as [19]. However, this investigation focuses on the tribology analysis of the contact of valve and valve seat ring. The dynamic analysis is mainly performed by experiments. A pursued target is to determine the location, force and load progression in the contact area between valve and valve seat ring. One of the findings is that the area of first contact between these bodies is only a fraction of the whole contact surface and therefore validates the assumption of a pointed contact of valve and valve seat ring during closing. The impact load on the valve is therefore much higher than suggested in former literature. The author points out that the cause of the eccentric but directional first contact depends on the valve train kinematic, especially in the direction of the rocker arm that transmits transverse forces to the valve. Furthermore, the author describes the valve closing and the ignition pressure load as two separate load events that have to be considered independently.

Another publications describing the valve secondary dynamic is provided by [29]. The empirical analysis of the guidance and the rocker arm pad shows that the tolerances can result in a valve misalignment and therefore in directional valve closing.

The thesis includes simulative FEM based analysis of the valve train mechanics with several detailing levels of the valve train. The loads and the valve secondary dynamic resulting either from valve seat ring misalignment or from the valve guidance tilting or from the overall valve train kinematic are analysed in their forms and effects on the first contact behaviour. The findings on the valve train mechanics build the basis for the tribological model. Furthermore, factors influencing the valve load and the tribology are deduced and included in the model. [4] present a dynamic analysis of the valve closing using an FEM model. The parametrisation of the model is performed with the aim to match the stress results of previously performed stress measurements.

An experimental approach on valve wear with focus on inlet valves is presented by [25]. This publication also considers aspects of the valve dynamic like the valve lift curve, the valve closing speed and the valve deflection caused by the gas pressure to determine the valve wear. [8, 24] point out the sliding occurring in the contact region between valve and valve seat ring during valve closing, which occurs after the first contact of valve and valve seat ring. This is an important mechanism considering wear. Because the valve closing has a high impact on the resulting valve loads, the sliding in contact also has to be considered for valve fatigue analysis. In general, publications with a focus on valve dynamic are rare. In contrast to this, the overall valve train dynamic is a more frequent object of research. The authors in [39] focus on the investigation of the valve clearance influence but not on the valve dynamic like the publications discussed before.

The authors use a FEM based model for the simulation of the valve train system. The components consist of either bars or beams, which allow a consideration of elastic deformation effects. The overall model has 20 DOF and uses spring-damper elements for contact modelling. The cam of the bottom-mounted camshaft is included in the model. Furthermore, the model contains a pushrod, a rocker arm, a valve, a valve spring, a valve guidance and a valve seat ring. With a small number of DOF simulation results can be generated really quickly.

[13] has a similar approach but includes the influence of a flexible camshaft on the overall valve train dynamic.

A further simulative investigation with focus on the overall valve train dynamic, especially on the influence of bearing misalignment is delivered by [1]. Further investigations on the valve train dynamic are carried out by [9]. This publication focuses on the dynamic of the valve train by simulating the valve train with a coupled flexible-rigid FEM model. This work also considers the behaviour of the valve train in the frequency domain, not only in the time domain.

The current approach at MAN ES regarding valve train dynamic simulation does not focus on the valve dynamic, either. Instead, the aim of the simulation is the analysis of cam contours. In this context, the results of interest are contact conditions between valve train components, the lifting and closing speed of the valves, the valve lift curves and whether or not a lift-off of components appears. The valves are rigid bodies in this model. They only have a realistic axial stiffness. The transverse stiffness in infinite, allowing no transverse movement of the valve. All contacts of the valve are simple spring-damper contact implementations. All further valve train bodies are also rigid bodies. The exceptions from that are the rocker arm, the valve bridge in case it exists in the valve train configuration and the cam follower. These components are flexible bodies. The simulation models at MAN ES including all valve train components are referred to as complete valvetrain (CVT). Figures 1.2 and 1.3 show the CVT models of the 49/60 and the 48/60 valve train.

The following subsections give a more detailed overview over the modelling of important components that are used in the above presented publications.

3.1.1 Valve Model

The simulation of the valve dynamic requires a flexible valve body. In addition, a flexible model delivers stress and strain information usable in a fatigue strength assessment.

There are different approaches of a flexible valve model. The FEM approach usually uses unreduced FE models, while the MBS approach uses models with a reduced number of DOF.

[19] uses a beam model for the MBS model that is able to consider valve bending and longitudinal expansion but does not consider the valve rotation. A mixed static-modal method is used for DOF reduction. The stiffness and mass matrix are obtained from the system matrices of the FE structure as shown in (2.1.39). The damping is obtained by a velocity-proportional damping model, without considering structural damping.

[26] uses computer-aided simulation to investigate the valve closing process. For this purpose, the thesis uses FEM based analyses of fine discretized volume models with corresponding material parameters.

Publications with focus on the overall valve train dynamic and not specifically on the valve dynamic use rigid body models to have a time efficient model with a small number of DOF. [39] use a valve model consisting of two bar elements with two nodes each as the focus is not on the valve dynamic but on the influence of valve clearance on the valve lift and the dynamic transmission error.

3.1.2 Valve Guidance Model

The valve shares a large contact area with the valve guidance. The guidance ensures a straight valve lift and reduces valve eccentricities. The contact area between the two bodies has no direct lubrication. Instead, lubrication oil from the top of the valve train is flows in the clearance between the guidance and the valve. It is therefore difficult to determine the height of the oil film.

The valve guidance in [19] is optionally rigid or elastic. The contact between the valve and the guidance considers the hydrodynamic lubrication reactions including the solid contact as explained in Sect. 2.1.4.2. The lubrication film reactions act as forces on one or several interface nodes on the valve body.

The analyses in [26] are FEM based. The simulation model has no specific implementation of the Reynolds equation and thus no implementation of hydrodynamic effects. Instead, the effects of the valve guidance lubrication regarding damping are considered by several beam elements applied to the valve shaft in radial direction. The guidance body is a rigid body consisting of shell elements. The length of the beam elements corresponds to the size of the guidance clearance. Therefore, also a solid contact is considered.

Models that include the primary valve dynamic but not the secondary dynamic like [39] only allow a vertical movement of the valve body. These implementation makes a valve guidance obsolete as the reaction forces and damping mainly occur in radial direction. Therefore, the model in this publication includes no guidance.

[20] points out the importance of a proper modelling of the interaction of valve and the valve guidance lubrication film. The inclusion of lubrication stiffness and damping enables the simulation of the valve secondary dynamic and therefore the analysis of the valve stresses resulting from eccentric valve closing in a subsequent FE analysis presented by the authors.

An efficient approach of the implementation of hydrodynamic interactions is presented by [21, 35]. These publications do not focus on the valve guidance but on crankshaft bearings. However, the time-efficient method of chart interpolation for the determination of structural loads on the joint is applicable for cases of cylindrical bearing geometry with negligible tilting and no special oil supply.

3.1.3 Valve Seat Ring Model

The depth of the valve seat ring modelling respectively the contact formulation between valve seat and valve seat ring depends on the main aim of the simulation. Most publications focusing on the valve train dynamic implement a stiff valve seat ring with a spring-damper contact force element as it is shown in [39]. This kind of implementation is a simplification and does not consider the impact of the conical geometry of valve seat ring and valve plate. The cones of valve plate and valve seat ring differ from each other by a few angular minutes to ensure the sealing of the closed contact. This geometry leads to a three-dimensional force vector on the real components.

As outlined in [19], a higher modelling depth is necessary for analyses focusing on the valve dynamic or valve seat ring wear. The implementation of the three-dimensional geometry with full contact surfaces of the valve seat ring enables the determination of contact forces for an eccentric contact of valve plate and valve seat ring.

The FE based analyses in [26] utilize a flexible model of the valve seat ring consisting of volume elements and therefore also consider the three-dimensional geometry of this component.

3.1.4 Valve Spring Model

The implementation of the valve spring model has a great influence on both the simulation time and also on the accuracy of the dynamic simulation. Therefore, the implementations of the valve springs differ significantly in various publications. The reason for this is that the spring dynamic has high influence on the overall valve train dynamic but also increases the simulation time a lot due to more detailed implementations.

[19] uses an uniaxial multi-mass spring model. The model consists of one mass element and two spring elements per coil. One of the spring elements is a linear element that represents the spring characteristics of the coil. The second spring element is non-linear and displays the coil-coil contact. Thus, local coil-coil contacts resulting from local spring dynamic are considered in the main spring axis direction. However, the point-to-point contact model of the coil-coil contacts leads to force peaks.

[3] present a valve spring modelling approach combining MBS and FEM. The simple valve train MBS model consisting of lumped masses and inertias connected by stiffness and damping values delivers the dynamic boundary conditions and loads for the detailed FE model of the valve springs. This model enables the evaluation of coil clashing and the resulting stresses in the valve spring.

[26] uses a spring consisting of beam elements. Beam elements are able to consider the transverse stiffness and dynamic of the valve springs. By using only one beam per spring, it is not possible to consider coil-coil contacts. The impact of multiple interacting masses is also not considered.

Reference [39] use the most simple but most time-efficient implementation of the spring following Hooke's law for the calculation of the resulting spring force and not considering further effects on the spring dynamic behaviour.

[10, 33] use a surge-mode approach to model the valve spring within a valve train model. Hence, the spring elements are described by the wave equation.

The approach presented by [11] focuses on the valve spring modelling and on the coil contact formulation in particular. A multi-mass model is used to model the spring. Each coil consists of lumped mass nodes. The coil contact is defined as uniaxial contact between those nodes of neighbouring coils. By omitting a penalty of the contact a more efficient solver algorithm can be applied to solve the coil contact of the valve spring.

3.1.5 Measurements of Valve Dynamic

[19] perform measurements for the purpose of simulation model verification. The setup is applied to a towed cylinder head. The major measurand is the valve velocity that is sensed by a laser-vibrometer. The valve lift and valve acceleration are determined by integration and differentiation, respectively. The valve secondary dynamic and the deviation of the valve stem in the valve guidance is detected by eddy-current sensors. The sensors are applied in two planes of the guidance. Each plane is sensed by two sensors with an angular distance of 90°.

Furthermore, strain gauges on the valve stem detect the valve stem force. The exact number of strain gauges is not specified, but the evaluation of these sensor signals only considers the tensile and pressure forces in the valve stem. The valve stem force gives several information on the valve dynamic. For example, the valve spring force acting on the valve stem can be detected. Moreover, the force indicates the valve load during valve closing.

[39] use the pushrod force and the measured pushrod strain to verify their model. The setup consists of a modified engine without crankshaft, piston and connecting rod. It is driven by a servo engine and therefore unfired. Strain gauges on the pushrod detect the pushrod strain. The comparison of predicted pushrod stresses and measured pushrod stresses shows a good accordance of measurement and simulation. These results verify the quality of the model regarding the relevant dynamic parameters. Also the increasing valve train dynamic with increasing rotational speed can be verified by the measurements.

In contrast to other publications using the measurements as experimental verification of simulation results, the analyses in [26] have a strong focus on experimental investigations. Especially the valve closing process is a subject of interest. The first investigation confirms the hypothesis of an eccentric valve closing. The valve closing direction is detected by strain gauges that are applied to a segmented valve seat ring. This setup only shows relative results as the valve seat ring has a thinned structure compared to the serial component. Furthermore, eddy-current sensors are applied in two planes of the guidance to detect the displacement of the valve stem in the guidance. The usage of the eddy-current sensors for the detection of the valve secondary dynamic is similar to the approach in [19]. This measurement assumes a stiff valve right before valve closing. The results of these investigations indicate a systematic eccentric valve closing. The magnitude of eccentricity depends on the components of the valve train drive and the rotational speed.

[4] present measurements results from strain gauges applied directly to the valve stem. The focus is the measurement of valve stem stresses during valve closing.

To determine absolute results of contact force between valve and valve seat ring during closing piezo-electric sensors are applied to the valve seat ring. These sensors do not change the seat ring stiffness significantly. This setup delivers a valuation basis of valve closing forces for various valve train geometries. The engine is towed. Therefore, the influence of ignition pressure is not considered.

An experimental approach specifically focused on the lubrication condition of the valve guidance and the interaction of valve and guidance oil film is given by [32]. The test up uses a glass cylinder and optical methods to measure the oil film distribution in the valve guidance for variable valve clearance and to determine the valve secondary dynamic movement in the guidance. The results show the benefits of a small clearance: The oil film is distributed evenly, resulting in less guidance wear.

[30] show valve dynamic measurements from a fired engine. The authors measure the valve lift and its derivatives with a magneto-resistive sensors sensor in the valve guidance. They also apply heat-resistant strain gauges to the valve stem measuring the valve load during engine operation. Due to the heat resistance of these sensors the influences of gas dynamic and temperature on the valve dynamic can be determined.

3.2 Valve Failure Evaluation

As pointed out before, the exhaust valve is a highly loaded component. Due to the high requirements regarding material strength also in a high temperature environment, the valve material selection is an important factor influencing the fatigue strength.

A common valve material is X45CrSi9-3. [34] analyses this material in the context of its application as valve material. Beside the bending stress and the gas load the publication points out axial stresses as a main load. Therefore it examines the axial fatigue strength of X45CrSi9-3.

The publications in the field of valve failure investigation mainly focus on specific failure incidents rather than on systematic failure mechanisms. There are both simulative approaches and also publications based on examinations of failed valves.

[36] investigates the fatigue failure of a passenger's car diesel engine exhaust valve. The investigations start by an examination of the failed component that shows typical beach marks of fatigue failure in the area of the critical fracture. Material defects and corrosion are excluded from possible failure mechanisms by the examination because it shows no indications of these. However, the main part of the analysis is carried out as FEM analysis.

For this analysis, three load cases of the valve are defined. The first load case is the force transmission from the valve spring to the valve stem, which has its peak value for the maximum spring deflection. The second load case is the non-uniformal temperature field with a maximum at the lower part of the valve stem between valve seat ring and valve guidance. The third load case is the combination of the first two load cases. The ignition pressure is not considered in these investigations. All load cases show small valve stresses in the critical stem area. Compared to the material's yield strength, the stresses resulting from one of the three load cases can not be the cause for the fatigue failure. From this, it is assumed that an additional phenomenon must be the main cause for the valve failure.

Further examinations of valves from the failed engine show carbon deposits on the valve plate. When added to the FE model, the carbon deposit leads to a bending stress in the valve stem. Under consideration of local material strength, depending on the valve temperature field, the area of the highest valve load relative to the local material strength corresponds to the area of failure of the real failed valve.

A further publication focussing on the thermomechanical valve loads is presented by [7]. The paper shows a numerical approach for a coupled thermomechanical analysis of the exhaust valve considering the operating valve temperature and the mechanical loads due to gas pressure in axial direction and the valve spring forces.

[23] present an analysis of the exhaust valve failure of a heavy-duty gas engine. The analysis is fully based on examinations of the failed valve that has already survived 5000 h of engine operation. Although the valve shows multiple flat fractures that are typical for fatigue, a systematic overload of the valve is excluded, as the failure would be expected after a few operating hours. Several observations indicate an overheating of the valve: A thick oxide layer is observed in the failed stem area, fretting is observed in the guidance contact area. The measurement of valve hardness show a much smaller value than expected for the material. The rapid decrease of the hardness is traced back to an overaging of the material induced by the overheating. The overaging of the material does not only lead to a reduced hardness but also to reduced material strength. The reduced material strength results in a fatigue failure of the valve within a short period of time. Also [17] detect an overheating of the exhaust valve as main reason for the fatigue failure of natural gas engine exhaust valve. The authors observed the significant decrease of hardness as well as an indicator of material overaging.

A reason of overheating can be found in the leakage of the exhaust valve. This common malfunction leads to an increasing temperature in the exhaust duct [22] and therefore on top of the valve plate.

Another metallurgical examination of a failed marine diesel engine valve presented by [31] shows, that the valve failure was caused by a insufficient hardening process of the valve face. Metallurgical examinations of failed exhaust valves from [6, 38] show, that the building of lamellar structures, the dissociation of the martensite phase or the coagulation of carbides in the valve material lead to the crack initiation and thermal yielding of the material, resulting in fatigue failure.

An investigation of the exhaust valve fatigue strength in form of a systematic analysis is presented by [27, 28]. In contrast to the previously introduced publications this study does not investigate the failure of single events that lead to the valve failure but obtains the durability of the valve under systematic influences.

The simulations are FE based. [27] uses all components from cam to valve. Both publications include the thermal influence on the valve material. In addition [27] consider the thermal deformations of valve guidance and valve seat ring resulting from the thermally induced cylinder head deformation. The resulting critical area at the lower part of the valve stem has the lowest safety factors and matches the findings from observed valve failure. The safety factor presented in the publication is the fatigue limit safety factor.

The effect on valve load due to the cylinder head deformation is also pointed out by [14]. The publication points out the resulting eccentricity between valve and valve seat ring due to the cylinder head deformation resulting in increased valve load.

A data-driven approach on the prediction of exhaust valve failure is presented by [12]. In contrast to previously outlined publications, this paper bases the prediction of exhaust valve failure events only on the statistical evaluation of engine operation data of 13 truck engines.

Another predictive approach considering the valve's operation conditions is presented by [5]. The valve fatigue prediction is based on a valve fatigue test apparatus. This machine is specified to apply a temperature distribution similar to the engine operation conditions to the exhaust valve to consider the decreasing yield strength of the valve material. Moreover, the test apparatus enables the application of variable axial load levels to the valve. However, the application of gas pressure is not possible. Although no bending load is applied to the valve in the test apparatus the fatigue test results outline a bending moment caused by dynamic effects during the valve actuation. As a result of this bending, one-sided valve seat wear is observed.

A further approach based on the building of a test apparatus is presented by [15]. In contrast to the previously presented approach, this test rig only considers tensile load by the valve spring and the impact during closing. The valve temperature is not considered. The test rig results are verified by a dynamic FEM simulation containing the valve, valve seat ring, grip jig and the rocker. The results are also compared to real failed valves. The failure location for the considered valves is on top of the valve stem instead of the bottom region, like in many other publications.

[16] present a simulative approach focussing on the valve impact as the main load event of the valve. The simulation of the stresses is FEM based. A predictive and simulative approach on exhaust valve fatigue analysis is presented by [29]. The FEM based analysis determines the valve stem stresses during closing. In this context a valve seat misalignment is considered. Furthermore, the temperature distribution of the exhaust valve is considered as a boundary condition in the analysis. The temperature distribution is determined in a previous computational fluid dynamics (CFD) analysis of the exhaust stream.

3.3 Fatigue Strength Assessment at MAN ES

This section focuses on the MAN ES approach of valve fatigue assessment. The approach is measurement based and also considers the statistical distribution of the resulting valve loads. This approach is the current state of the art at MAN ES, specifically considering heavy-duty engine valves.

The measurement approach developed in the following chapter of this thesis is based on the current measurement approach at MAN ES. Moreover, evaluations of the measurement data from the current approach deliver information on the valve dynamic (see Sect. 4.1).

The MAN ES approach of valve design assessment has the aim to compare the measured valve load data to a reference value. The reference value is an empirical value of valve strain. The empirical value is based on measurements on engines with failed valves but with an added safety margin. The strain measurement position is the horizontal plane at the middle of the valve radius. The valve radius middle is assumed to be the critical area regarding fatigue, although there are publications that indicate the critical valve area to be not in the middle of the radius, but in the upper region of the radius or in the lower region of the valve stem, respectively. Due to the size of the heavy-duty engine valves, it is possible to apply several sensors directly to the valve. Therefore, no further modifications of guidance and cylinder head are necessary. Four strain gauges are applied to determine bending strain around the circumference. The measurement cycle is repeated multiple times to determine the 99.9% quantile of maximum strain. This value is then compared to the empirical limit.

The valve temperature during operation is also obtained by measurements and compared to empirical limit values as well. The appearing temperature data and strain data are not related to each other. Instead, the current approach assesses temperatures and strain data separately. The measurements are performed on test bed engines. These are fired engines operating in all common engine operation points. Several engine types are available for those measurements. There are two categories of test bed engines: There are single cylinder engines, which are used for research and development. Furthermore, there are serial engines with six to twenty cylinders. Because of the comparison to empirical values each change in the engine operation conditions or valve train design requires a new assessment of the valve fatigue strength, because the absolute strain value changes for each modification.

The sensors applied directly to the valve require cables that transmit the measurement signals. These cables are led out of the cylinder head and require the valve to not rotate during operation. A rotation of the valve would twist the cables and destroy the measurement setup.

3.3.1 Strain Measurements

The assessment of valve fatigue strength by strain measurement is combined with further measurements on the valve train.

Another measurement in this context is the measurement of the valve lift. The measurement of the valve lift enables the assignment of specific valve load phase to the valve lift position. This position measurement is either carried out with inductive displacement transducers for older measurement setups or with an optical position sensor that is more accurate for newer measurements.

The measurement of the cylinder pressure via a pressure sensor determines the ignition pressure characteristics. Further strain gauges on the pushrod and on the valve springs measure the strain of these components that are transformed into force signals. This measurement setup is also called CVT measurement.

The measurement position of valve strain is the same for all valve geometries, although some geometries have a hollow groove in the radius area and some valve geometries have a thicker conic shape on the lower stem part. The lower stem of the

Fig. 3.1 Position of strain sensors on valve

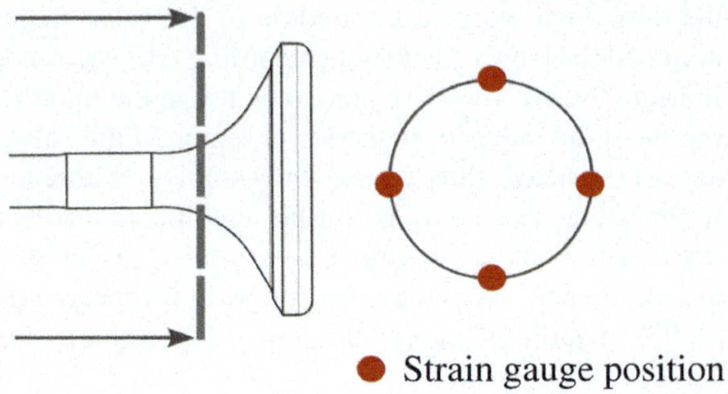

valve is in direct contact with the hot exhaust stream causing a decreasing material strength.

Four strain gauges are placed in the horizontal plane at the valve plate radius with a 90° circumferential distance to each other. Figure 3.1 shows this setup. A similar measurement approach on the valve stem strain with high-temperature strain gauges is presented by [37].

In fact, three strain gauges are sufficient to determine the maximum circumferential strain. However, the failure rate of the strain gauges is increased at the valve radius due to the high temperature. Therefore, the application of four strain gauges provides additional robustness in the setup.

The evaluation of the strain data, shown in Sect. 3.3.3, determines the strain components in dependency of the angle of the valve radius circumference. Therefore, the strain for each position in the measurement plane can be determined. The strain values are multiplied with the steel Young's modulus to generate more practical stress values. It needs to be pointed out that these stress values do not comply with real stresses, because the thermal variability of the Young's modulus is not considered. As this operation does not deliver real stress values the multiplication by the Young's modulus is not used for further considerations in this thesis. Instead, only strain values are considered.

The accuracy of the empirical reference value, which the experimental valve strain values are compared to, is questionable as there are valve strain data from engines without damages exceeding this reference value limit. An example of valve measurements with limit exceeding strain data is shown in Fig. 3.2. For this engine type there has been no fatigue-related failure of valves. This observation on a valve train of a serial engine indicates the inaccuracy of the empirical value. It is unclear which consequences the exceeding of the empirical value has and when a fatigue-related failure occurs.

In the following, the valve train measurements are explained in more detail.

There are three main types of valve train measurements. The first type is the multi segment measurement (MSM). This measurement obtains the valve strain in the valve plate radius during its main load phases with high resolution. Multiple measurement segments of the same engine operation point are generated to obtain the distribution of maximum valve strain. The resulting strain data is compared to the

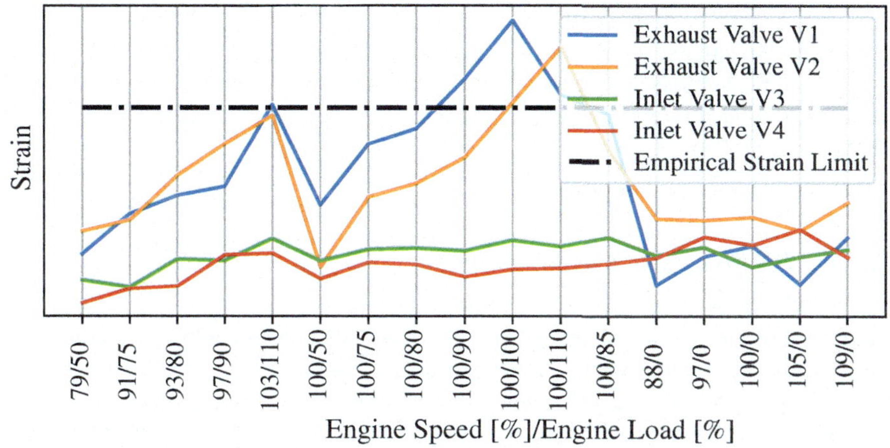

Fig. 3.2 Maximum strain of 99.9% quantile

empirical limit of valve load. However, a weakness of these methods is the lack of predictivity because a new valve design needs to be fully released and manufactured to be tested. This method considers the physical backgrounds of the valve dynamic only roughly.

The evaluation of measurement data queries the empirical reference values since few data show an exceeding of the reference value. Moreover, the empirical values are often deduced from older engine types and therefore from older measurements. Parameters, which might influence the outcome of the measurement like the specific engine load, the engine operation temperatures or the valve material are not considered within the empirical value. Furthermore, the strain measurements do not consider thermal influence.

Another measurement type obtains the valve strain in the middle of the stem. Due to the smaller space in this area, only two strain sensors are applied in the measurement plane. Hence, the strain can only be obtained at the sensor positions and provide information on the bending vibration amplitude of the valve. Due to the sensor position on the stem, the valve guidance has to be modified to avoid the contact of the sensors with the guidance. Thus, the guidance clearance in the middle part of the guidance, which is the potential contact area between strain gauge and guidance, is increased. This prevents the strain gauges from getting into contact with the guidance during engine operation. The actual guiding areas of the guidance are therefore reduced to the upper and the lower section of the valve. This setup influences the dynamic of the valve.

The third type of measurement is the single segment measurement (SSM). The measurands are the same as in the MSM. However, measurands like the valve spring forces and the pushrod forces are added to obtain information on the overall valve train dynamic. The valve train dynamic is the focus of this measurement type. The valve load can not be determined accurately, because the resolution of all measurand channels is lower compared to the MSM. The load peaks of the valve might be filtered by the smaller resolution. The reason for the smaller resolution is the higher number of measurands. There is a total resolution for the measurement split up by all measurands. The more measurands there are, the lower the resolution per measurand channel.

3.3.2 Temperature Measurements

Temperature measurements are carried out on exhaust valves only, since they experience much higher temperatures than the inlet valves. In most cases, only the valve plate bottom site and the valve seat are measured as they are assumed to be the hottest areas of the valve. Four temperature sensors of type NiCr-Ni thermocouples are positioned in these areas, one sensor is placed in the middle of the valve plate bottom site, the other sensors are distributed over the valve seat. A similar approach on the measurement of valve plate temperatures is presented by [37]. The measured temperature values are compared to reference values for evaluation. There are no further evaluation steps.

The temperature measurements are not synchronized to the strain measurements regarding considered engine operation points.

3.3.3 Strain Measurement Evaluation

This subsection describes the evaluation of strain measurements carried out in the valve plate radius. The four strain signals of the valve radius enable a calculation of the overall strain maximum in the radius. The equations in (3.3.1) provide strain components in Cartesian coordinates. The axial valve direction is the y-direction with the positive direction pointing to the valve stem top face. The x-direction points to the liner centre, the z-direction results from the definition of x- and y-direction in the right-handed Cartesian coordinate system.

$$\epsilon_x = \frac{\epsilon_{0^\circ} - \epsilon_{180^\circ}}{2} \tag{3.3.1a}$$

$$\epsilon_y = \frac{\epsilon_{0^\circ} + \epsilon_{90^\circ} + \epsilon_{180^\circ} + \epsilon_{270^\circ}}{4} \tag{3.3.1b}$$

$$\epsilon_z = \frac{\epsilon_{90^\circ} - \epsilon_{270^\circ}}{2} \tag{3.3.1c}$$

These are used to analyse the strain as a function of φ, the angular position on the valve radius circumference (3.3.2). The strain maximum can be derived from this. This determination of the maximum strain evaluates the MSM strain measurement data.

Each MSM segment delivers a maximum strain. By assuming a normal distribution of the strain maxima the strain value corresponding to the 99.9% probability serves to compare with the empirical strain limit.

$$\epsilon_{\varphi} = \epsilon_y + \epsilon_x \cdot \cos\phi + \epsilon_z \cdot \sin\phi \tag{3.3.2}$$

$$\epsilon_{max} = \max(\epsilon_{\varphi}) \tag{3.3.3}$$

The normal distribution applies to many applications (see Sect. 2.2.1.1). Therefore, the assumption of a normal distribution of the strain maxima appears obvious. However, the distribution of the measured strain maxima differs from the normal distribution significantly. The deviations from normal distribution of the affected measurements do not follow a specific scheme. The form of the maximum strain distribution rather shows various forms and can therefore hardly be described by any other distribution. Therefore, the calculated 99.9% probability level of the maximum strain is not reliable, because the real distribution and its parameters are unknown. The real failure probability can not be determined.

The evaluation of the SSM delivers no fatigue strength assessment relevant data. However, the evaluation of the side measurands like the spring forces and pushrod forces indicate the behaviour of the valve train dynamic. The following chapter shows more details of these evaluations.

3.4 Summary of the State of the Art

The chapter showed the state of the art regarding valve dynamic simulation and valve fatigue strength assessment.

Regarding the simulation of the valve train, the FEM method is common, but requires simplifications of clearances and contact formulations. The FE models often do not simulate the full valve working cycle.

The MBS approach does not need these simplifications and is therefore used when detailed valve dynamic investigation is the main investigation aim.

There are publications evaluating the valve train dynamic but focus on other components like the cam contour. These investigations usually use a simplified valve model. The publications focusing on the valve dynamic use a more detailed model of the valve, mostly a flexible body model.

There are publications considering the eccentric valve closing. The valve seat ring misalignment, wear effects, guidance tilting and cylinder head deformation are considered causes of this eccentricity. Further external influences by deformations or forces are not considered.

Several publications use measurements to validate the simulation model. All measurements are carried out on a trailed, unfired engine. The influences of temperature and ignition pressure are therefore not considered.

Regarding the fatigue assessment of the valve, there are publications evaluating failure events of valves either by simulation or by examination. There is also an existing method on a systemic fatigue limit assessment of the valve. The loads for this fatigue limit assessment are determined by an FE model.

The present thesis mainly uses MBS to analyse the valve dynamic which accords with the current state of the art. In contrast to other MBS models, this thesis uses an OVM and leaves out rocker arm, pushrod, cam follower and cam. Instead of including these components, variations in valve actuation are applied by a variation of the model boundary conditions.

The investigations on the valve dynamic and valve loads consider component deformations like the temperature induced cylinder head deformation or a valve seat ring deformation or misalignment. However, the field of considered influences is increased, compared to the state of the art, to further systemic influences on the valve. These influences include external forces and component deformation and misalignment, e.g. the misalignment of the valve spring. This thesis considers not only the eccentric valve closing, but investigates to valve dynamic during the complete valve lift phase.

The measurements in this thesis are carried out on a fired engine, and hence include temperature and ignition pressure influence. As a result, these measurement conditions are closer to real engine operation conditions and provide a more reliable data base. Moreover, the strain of the valve can be measured due to the large size of heavy-duty engine valves.

The fatigue strength assessment follows an analytical approach of the FKM guideline and uses systemic loads. Therefore, this fatigue assessment method is transferable to other engine types with different loads and different component geometries. In contrast to analyses of single failure events, this approach delivers a predictive fatigue strength assessment. The material-specific strength values including temperature influences are considered within the fatigue strength assessment. The load side considers statistical influences on the valve load. The statistical analysis of valve loads delivers new findings on the distribution of valve loads. Moreover, the variances of several valve train parameters influencing the valve load are considered. This statistical analysis of valve loads and parameter influences delivers a method to determine a valve load collective purely based on simulation data. This increases the frontloading of the valve design process compared to current approaches.

References

1. Boucke, A. ; Lang, J. ; Knoll, Prof. Dr. - Ing. G.; Ali, M. Y. ; Ejakov, M. : Untersuchung der Auswirkung von Einbauabweichungen von Ventiltriebskomponenten auf Ventilführung und Sitz mittels numerischer Simulation: Anwendung am Beispiel eines Antriebs per Rollenschlephebel in Zug- und Schubrichtung von Ford Motor Company. In: *Ventiltrieb und Zylinderkopf 2017* (2017), 159–172
2. Braess, D. : *Finite Elemente: Theorie, schnelle Löser und Anwendungen in der Elastizitätstheorie*. 4. Berlin, Heidelberg : Springer Berlin Heidelberg, 2007
3. Calabretta, M. ; Sitta, A. ; Oliveri, S. M. ; Sequenzia, G. : Simulation of Dynamic Stresses on High Performance Engine Valve Spring System Considering Coil Clashing Effect. In: *Advances on Mechanics, Design Engineering and Manufacturing III* (2021), 108–113
4. Cavalieri, F. J. ; Luengo, C. ; Risso, J. ; Zenklusen, F. ; Cardona, A. : Numerical and Experimental Stress Analysis of an Internal Combustion Engine Valve During the Closing Event. In: *Proceedings of the Institution of Mechanical Engineers, Part D: Journal of Automobile Engineering* 228 (2014), Nr. 5, 479–489
5. Choi, J. ; Lee, J. ; Jun, N. ; Seok, C.-S. ; Park, S. ; Kim, G. : Development of Laboratory Fatigue Testing Apparatus for Automotive Vehicle Engine Valve Simulating Actual Operating Conditions. In: *International Journal of Precision Engineering and Manufacturing* 20, Nr. 7, 1241–1253
6. El- Bitar, T. ; El- Meligy, M. ; Khedr, M. : Investigation of exhaust valve failure in a marine diesel engine. In: *Engineering Failure Analysis* 114 (2020), 104574.
7. Fersaoui, B. ; Cerdoun, M. ; May, A. ; Carcasci, C. : Thermo-mechanical stress analysis within a steel exhaust valve of an internal combustion engine. In: *Proceedings of the Institution of Mechanical Engineers, Part C: Journal of Mechanical Engineering Science* 236 (2022), Nr. 1, 635–654
8. Forsberg, P. ; Gustavsson, F. ; Hollman, P. ; Jacobson, S. : Comparison and analysis of protective tribofilms found on heavy duty exhaust valves from field service and made in a test rig. In: *Wear* 302 (2013), Nr. 1, 1351–1359
9. Guo, J. ; Zhang, W. ; Zhang, X. ; Cao, Y. : Dynamic and exciting analysis with modal characteristics for valve train using a flexible model. In: *Mechanism and Machine Theory* 78 (2014), 158-176.
10. Guo, J. ; Zhang, W. ; Zou, D. : Investigation of dynamic characteristics of a valve train system. In: *Mechanism Machine Theory* 46, Nr. 12, 1950–1969
11. Haslinger, J. ; Offner, G. ; Sopouch, M. : Non-Smooth Dynamics of Coil Contact in Valve Springs. In: *Zeitschrift für Angewandte Mathematik und Mechanik* 94 (2014), Nr. 11, 957–967.
12. He, Y. ; Kusiak, A. ; Ouyang, T. ; Teng, W. : Data-driven modeling of truck engine exhaust valve failures: A case study. In: *Journal of Mechanical Science and Technology* 31, Nr. 6, 2747–2757
13. Hu, B. ; Li, Y. ; Yin, L. : Theoretical and Experimental Analysis of Dynamic Characteristics for a Valve Train System. In: *Sensors* 21 (2021), Nr. 19.
14. Huber, C. O. ; Zurk, A. : Cylinder Head Solutions for 300 bar and beyond: Next generation commercial and large engines advanced design and simulation. In: *VDI-Fachtagung Ventiltrieb und Zylinderkopf 2019* (2019), 171–190
15. Jun, N. ; Park, J. ; Kim, K. ; Park, K. ; Seok, C.-S. : Method for evaluating the fatigue life of stem end considering the structural characteristics of engine valve for automobiles. In: *International Journal of Automotive Technology* 23, Nr. 2, 303–313
16. Jun, N. ; Seok, C.-S. ; Park, K. : Evaluation of Fatigue Life of Automotive Engine Valves Considering the Dynamic Behavior. In: *International Journal of Automotive Technology* 23, Nr. 6, 1717–1726
17. Khan, M. A. ; Khan, M. I. ; Shakoor, D. A.: A failure analysis of the exhaust valve from a heavy duty natural gas engine. In: *Engineering Failure Analysis* 85 (2017), 12.
18. Klein, B. : *FEM: Grundlagen und Anwendungen der Finite-Elemente-Methode im Maschinen- und Fahrzeugbau*. Wiesbaden : Springer Fachmedien Wiesbaden, 2015

19. Knoll, G. ; Brands, C. : Forschungsvorhaben 614: Rechnergestützte Simulation der Dynamik und der Beanspruchung des Ventiltriebs. In: *FVV Forschungsheft* 661 (1998)
20. Knoll, G. ; Brands, C. ; Schönen, R. : Elastohydrodynamische Kopplung von Nockenwelle und Lagerung unter Berücksichtigung der Ventiltriebsanregung. In: *MTZ - Motortechnische Zeitschrift* 64, Nr. 9, 728–734
21. Knoll, G. ; Schönen, R. ; Wilhelm, K. : (1997)
22. Kowalski, J. : An Experimental Study of Emission and Combustion Characteristics of Marine Diesel Engine in Case of Cylinder Valves Leakage. In: *Polish Maritime Research* 22 (2015), Nr. 3, 90–98.
23. Kwon, O. G. ; Han, M. S.: Failure analysis of the exhaust valve stem from a Waukesha P9390 GSI gas engine. In: *Engineering Failure Analysis* 11 (2004), Nr. 3, 439–447.
24. Lewis, R. ; Dwyer- Joyce, R. : *Automotive Engine Valve Recession*. Professional Engineering Publishing, 2001 (Engineering Research Series (REP))
25. Lewis, R. ; Dwyer- Joyce, R. : An experimental approach to solving combustion engine valve and seat wear problems. In: *Tribology Series* Bd. 39. Elsevier, 2001, 629–640
26. Messer, M. : *Tribologische Untersuchungen des Systems Ventilsitz - Ventilsitzring*. Hannover, Gottfried Wilhelm Leibniz Universität, Dissertation, 2013
27. Oh, K.-C. ; Cho, S.-W. ; Kim, J.-H. : A Study of Durability Analysis Methodology for Engine Valve Considering Head Thermal Deformation and Dynamic Behavior, 2014
28. Pang, M. L. ; Smith, S. P. ; Herman, R. ; Buuck, B. : Stress analysis of an automotive engine valve by finite element methods. In: *SAE Technical Paper Series*, SAE International
29. Roth, G. : Fatigue analysis methodology for predicting engine valve life. In: *SAE Technical Paper Series*. Warrendale, PA, United States : SAE International, März 2003
30. Schwarz, D. ; Bach, M. ; Fuoss, K. : Messungen der Ventiltriebsdynamik am befeuerten Verbrennungsmotor. In: *MTZ - Motortechnische Zeitschrift* 70, Nr. 6, 480–487
31. Smoleńska, H. ; Kończewicz, W. : Failure analysis of the exhaust valve face in diesel marine engine. In: *Advances in Materials Science* 10 (2010), Nr. 2, 11–18.
32. Todsen, U. ; Herbst, O. ; Krüger, G. ; Biallas, E. : Optical ways to improve the tribology of valve guides. In: *SAE Technical Paper Series*, SAE International
33. Vasilyev, A. ; Bakhracheva, Y. ; Storojakov, S. : Dynamics Simulation Model for the Internal Combustion Engine Valve Gear. In: *Procedia Engineering* 150 (2016), 312–317
34. Voorwald, H. ; Coisse, R. ; Cioffi, M. : Fatigue Strength of X45CrSi93 stainless steel applied as internal combustion engine valves. In: *Procedia Engineering* 10 (2011), 1256–1261
35. Warmuth, O. ; Brohmer, A. ; Schönen, R. ; Träbing, C. : Transiente Analyse der Kurbelwellenbeanspruchung unter Berücksichtigung der nichtlinearen Schmierfilmsteifigkeit in den Grundlagern. In: *Reibung, Schmierung und Verschleiß, Forschung und praktische Anwendungen, Tribologie-Fachtagung, 2002*. Gesellschaft für Tribologie
36. Witek, L.: Failure and thermo-mechanical stress analysis of the exhaust valve of diesel engine. In: *Engineering Failure Analysis* (2016), Nr. 66, 154–165
37. Worthen, R. P. ; Rauen, D. G.: Measurement of valve temperatures and strain in a firing engine. In: *SAE Technical Paper Series*. Warrendale, PA, United States : SAE International, Febr. 1986
38. Yu, Z. ; Xu, X. : Failure analysis and metallurgical investigation of diesel engine exhaust valves. In: *Engineering Failure Analysis* 13 (2006), Nr. 4, 673-682.
39. Zhou, C. ; Hu, B. ; Chen, S. ; He, L. : An enhanced flexible dynamic model and experimental verification for a valve train with clearance and multi-directional deformations. In: *Journal of Sound and Vibration* (2017), Nr. 410, 249–268

Chapter 4
Valve Dynamic Investigations and Measurements

The former chapters described the main aim of this thesis, the theoretical background of the methods used to perform the necessary analyses and the measurement methods, and contextualized these to the current state of the art. The current chapter shows the new investigations of the valve train dynamic and the valve dynamic in particular. The analyses are simulation based. Therefore, it is important to ensure that the simulation model shows a good accordance with the real valve dynamic. The validation of the simulation model is therefore based on measurements on the valve train. These measurements and their evaluation are shown in the first part of the chapter.

The second part of this chapter describes the simulation model used for all dynamic analyses in this thesis. Then, the analyses performed to determine the cause of the valve secondary dynamic are shown. Finally, measurement and simulation results are compared to determine the accuracy of the final simulation model.

4.1 Measurements

This section shows the development of new methods for the evaluation of valve train measurements. The existing approach on the statistical evaluation has deficiencies regarding the assumed normal distribution of valve strain maxima, as Sect. 3.3.3 pointed out. However, to generate information from the huge amount of valve strain data that has been generated prior to this thesis, this section starts with the development of improved methods on the evaluation of valve strain measurements. Findings from this process are used to define an improved measurement setup for new valve strain measurements.

In addition, the section describes the setup, execution and the evaluation of the measurements later used for the purpose of simulation model validation. The test bed for these measurements is a single cylinder engine of type MAN ES 49/60 running in the diesel process. The large valve size of the 49/60 engine makes this engine type suitable for strain measurements. The 49/60 valve train has the largest valves

A. Eret, *Dynamic and Fatigue Assessment of Heavy-Duty Engine Valves*, Mechanics and Adaptronics, https://doi.org/10.1007/978-3-031-49151-1_4

of the available test bed engines which makes it possible to apply more sensors on the valve without influencing the mechanical characteristic of the valve significantly. With strain sensors applied in the valve stem middle and in the valve plate radius the measurement delivers both information on the valve dynamic and data for the determination of valve load data.

Moreover, there are separate valves with temperature sensors, because the exhaust valves reach temperatures up to 500 °C. As a result, significant decreasing in local material strength is expected. Thus, the temperature data has to be included in the fatigue strength assessment.

4.1.1 Development of Statistical Methods of Valve Strain Data Evaluation

The description of former measurements and its evaluation in Sect. 3.3 shows the problem of the assumed normal distribution not describing the distribution of the maximum strain of valve closing accurately. In addition, the assumption of the valve closing to be the highest load during the working cycle is deduced from measurement observations, but has not been systematically shown for different engine types, yet. Thus, this section shows the systematic analysis of valve load phases. Further investigations attempt to find correlations of the maximum strain amplitude and engine operation parameters like rotational speed or engine load.

The development of this method of statistical evaluation is based on measurement data generated prior to this thesis. The explanations in this section refer to and complement the ones in [5].

4.1.1.1 Determination of the Dominant Valve Load Phase

The determination of the maximum load phase utilizes the SSM because they include at least two full working cycles. The potentially critical phases shown in Fig. 4.1 of the valve are:

- Ignition (blue)
- Valve Opening (green)
- Valve Closing (red).

The analysis includes ten different engine types. It compares the amplitudes of the valve strain during closing to either the valve strain during opening or the strain during ignition pressure for each valve and engine operation point. The exemplary result of a 45/60 engine in Fig. 4.2a and b shows the overall result of the analysis. The x-axis shows the maximum valve load during ignition (Fig. 4.2a) or the maximum valve load during valve opening (Fig. 4.2b), while the y-axis shows the maximum valve load during valve closing. The black dashed line marks the line of equivalent strain of

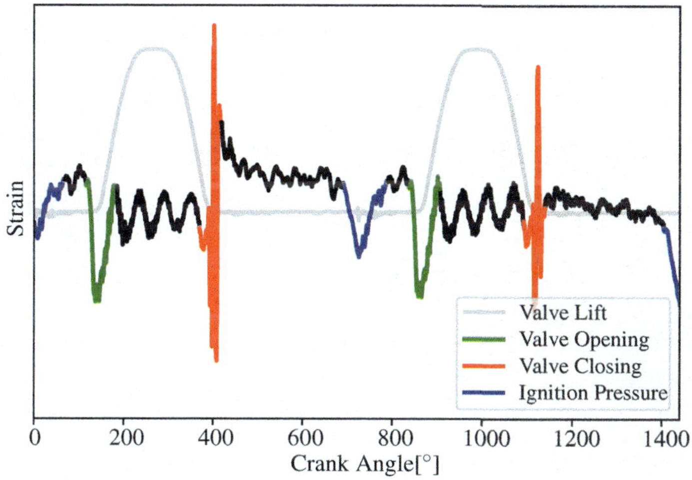

Fig. 4.1 Example of valve load phases

the compared load phases. This means that an operation point lying on the dashed line has the same maximum load during the valve closing and valve opening or ignition, respectively. Operation points above the dashed line have a higher maximum load during valve closing than during valve opening or during ignition.

Both figures show that the majority of all operation points has a higher maximum load during valve closing than during the compared phase. Only a few engine operation points indicate a higher valve load during valve opening or ignition, respectively. In addition, the operation points not having their maximum valve load during valve closing are always operation points of generally lower valve load. That means that the valve closing load is not only higher compared to the other load phases during the same engine operation point but the valve closing load also has the highest amplitudes considering all engine operation points. The shown results can be reproduced for several engine types with different configurations.

4.1.1.2 Determination of Valve Load Distribution

Section 3.3 already pointed out the fact that the strain maxima distribution is sometimes poorly described by the normal distribution. This finding is emphasised by probability plots in Fig. 4.3 showing the distribution of measured strain maxima. The x-axis of these probability plots shows the maximum strain amplitude of one segment while the y-axis shows the cumulative frequency of the results. The axis is logarithmic scaled which leads to a linear curve of the normal distribution. In the context of distributions, the y-axis is to be interpreted as the probability. The assumption of a normal distribution applies to several engine operation points on several engines.

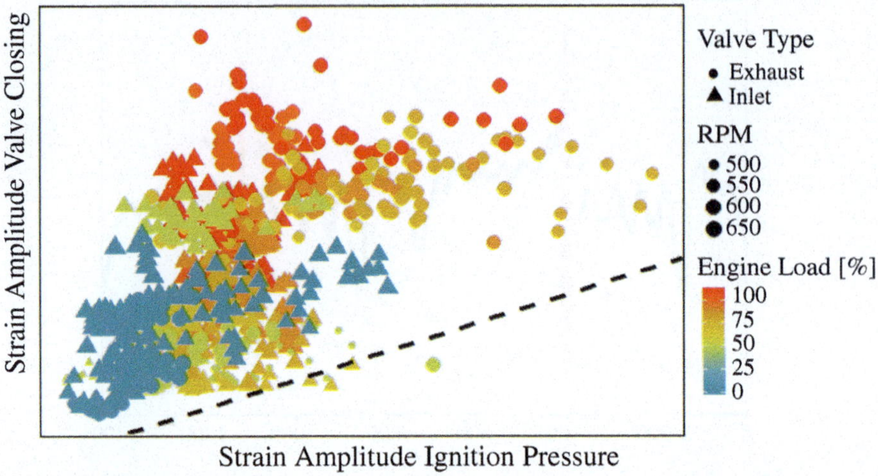

(a) Comparison of maximum strain amplitudes, valve closing vs. ignition pressure

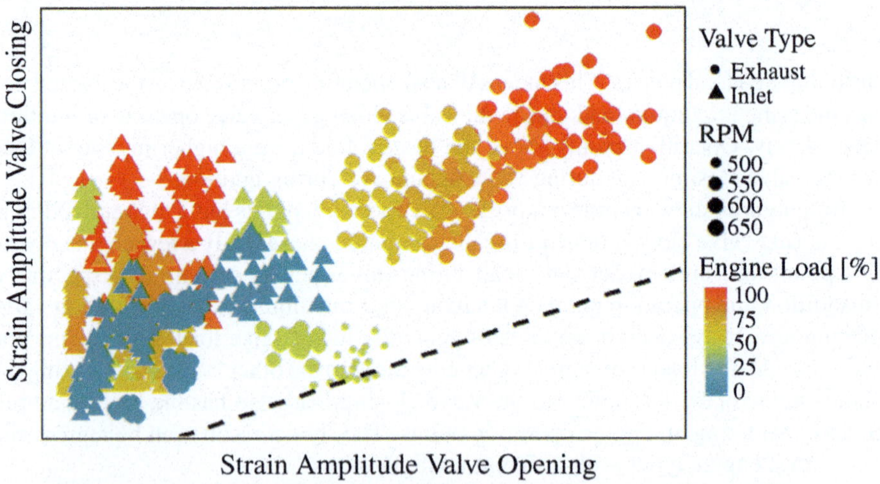

(b) Comparison of maximum strain amplitudes, valve closing vs. valve opening

Fig. 4.2 Determination of maximum valve load phase

However, there is a non-negligible number of engine operation points that cannot sufficiently be described by the normal distribution. Figure 4.3a and b show examples of such strain distributions. The deviations from the normal distribution are not consistent, but show various distribution forms. For some measurements, the maximum strain is lower than the strain of the normal distribution for a specific probability. This makes the assumption of a normal distribution a conservative assumption that potentially leads to higher costs due to bigger component dimensions or more expensive materials with higher material strength.

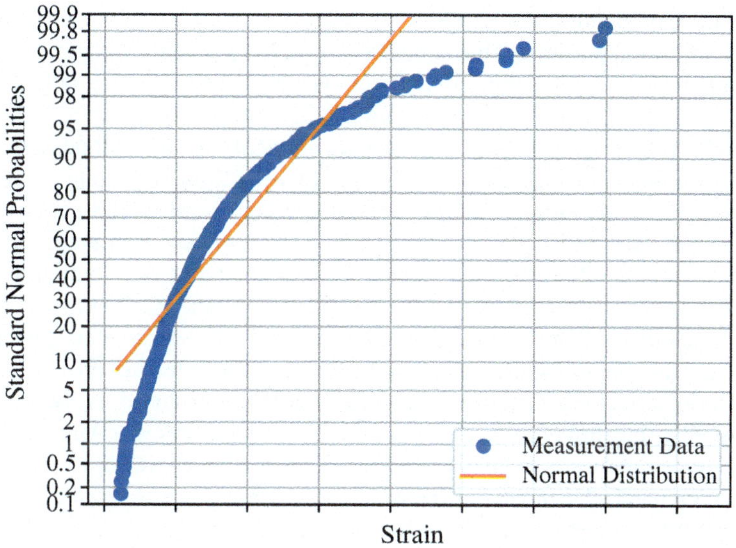

(a) Distribution of 49/60 strain data on exhaust valve, 110% Load/600 rpm, normal distribution overestimates valve load

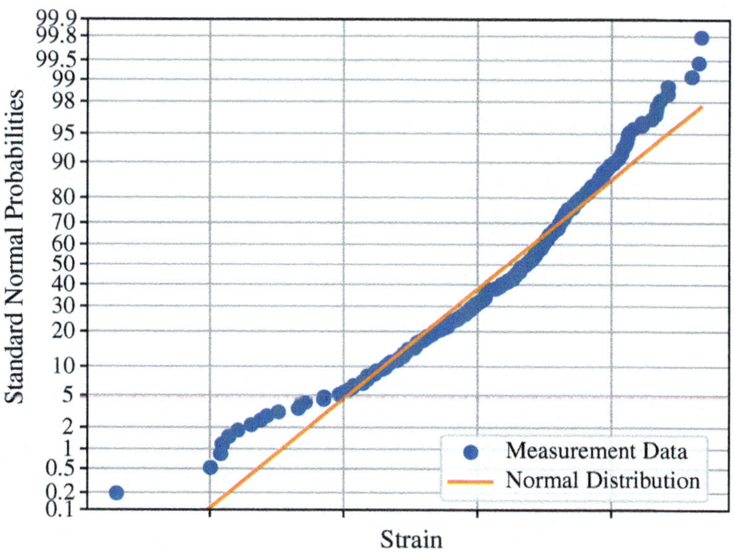

(b) Distribution of 32/44 strain data on exhaust valve, 110% Load/750 rpm, normal distribution underestimates valve load

Fig. 4.3 Distributions of valve strain data

Other measurements show a higher maximum strain load than the corresponding normal distribution predicts. This is the more problematic case, because the normal distribution underestimates the actual strain load and therefore potentially leads to a valve design that is likely to fail under the real operational load.

Because of the various forms of strain data distributions, it is not possible to find a distribution that is able to describe all strain distributions for all engine operation points. Regarding the fatigue strength assessment this is indeed not necessary. The important part of the distribution is the part of the highest strain amplitudes with the highest cumulated probability. Hence, the extreme value distribution described in Sect. 2.2.1.1 describes this part of the measurement data distribution.

For the application of the approximation in (2.2.5), the parameters ξ, $\tilde{\sigma}$ have to be estimated and the threshold u_{tr} has to be chosen. The selection process is shown in [5]. The method of estimation selected to determine ξ and $\tilde{\sigma}$ is the method of moments. The threshold is the 95% quantile of the measurement data. On the one hand, the threshold needs to be high enough to ensure the applicability of the GPD. On the other hand, the threshold must not be too high to ensure that the sample size is still high enough, because a threshold of 95% means that only 5% of the maximum data is used for the fitting of the distribution. The threshold is determined by evaluations of older measurement data. The resulting distributions with either a threshold of 90 or 95% are compared to the measured maximum. The 95% threshold based distribution shows realistic estimations of the maximum strain while the 90% threshold based distribution even assumes a lower strain maximum than the measured strain maximum for some data sets.

4.1.1.3 Influences on Maximum Strain

The measurements of valve strain include several engine operation points. The following variance analysis investigates whether the valve strain depends on either engine load or engine speed. The analysis is performed for several engine types. The influence of the load is determined by evaluating the strain path over varying load for constant engine speed. The analysis is repeated for different levels of engine speed. The influence of the engine speed is determined by the same method, but with constant load level and varying engine speed. The approach is shown in [5]. Figures 4.4 and 4.5 show results of the investigation. Two findings can be derived from that analysis: The first finding is that the variance in engine types and engine operation points leads to a small number of engine operation points with a constant engine speed for varying engine load and vice versa. As a result, it is not possible to analyse these influences on all engines. The second finding for engines where the correlation analysis is applicable is that there are no clear results. Figure 4.4 shows the results for constant load of a 51/60 exhaust valve. For engine loads of 0% the strain amplitude increases with increasing engine speed. For 110% engine load the strain decreases with increasing engine speed. Similar results can be obtained for constant engine speeds and variable loads. Figure 4.5 shows the analysis for a 32/44 exhaust valve. For an engine speed of 720 rpm the strain increases with increasing load, for

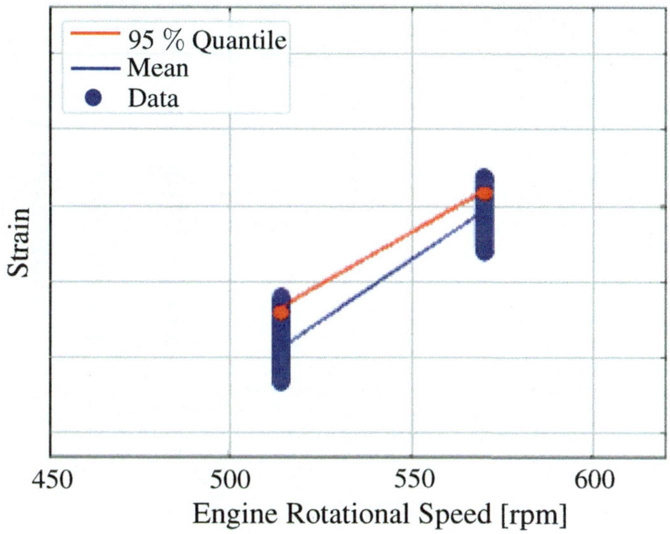

(a) 0% Load

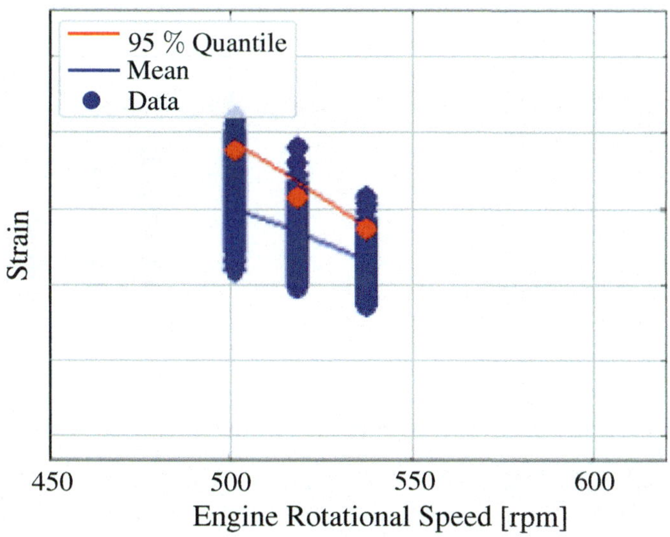

(b) 110% Load

Fig. 4.4 Influence of engine speed on strain, 51/60 engine

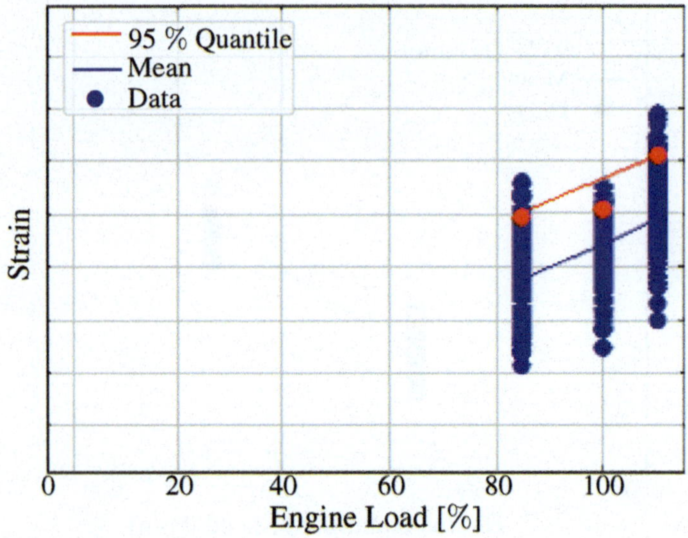

(a) 720 RPM

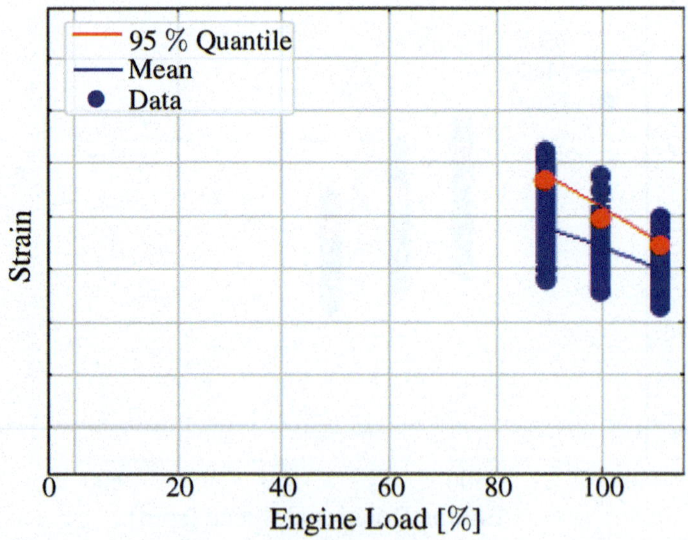

(b) 750 RPM

Fig. 4.5 Influence of engine load on strain, 32/44 engine an engine speed of 750 rpm the strain shows an opposite behaviour for increasing engine load. The results indicate that engine load and engine speed have no linear influence on the maximum valve strain. The engine speed and engine load might also interact with other influencing factors that are not captured by the measurements.

4.1.2 Setup

On basis of all the findings from former measurements a new measurement setup is derived within the scope of this thesis. The following subsection describes the measurement setups for the temperature measurements as well as for the valve strain and dynamic measurements. Regarding sensor application there are two cylinder heads equipped, one with the setup for temperature measurements, the other one with the setup for strain and dynamic measurements. The usage of separate cylinder heads has the benefit of a fast test bed conversion from temperature to dynamics and strain measurement. The application of temperature sensors and strain sensors to the same valve is not possible, because the space on the valve is limited. Altogether 14 sensors are necessary. Each of these sensors needs an own cable groove. The number of notches on a single valve would influence the mechanical characteristic of the valve too much. Therefore, the measurements are split into two separate measurement setups. Excluding the measuring components that are the modified valves, the slightly modified guidance, the axial bearing replacing the rotocap in the measurements and the rotation lock all components of the cylinder head assembly correspond to the standard 49/60 dual fuel cylinder head. Therefore, the operating conditions of the measured valve train and the serial valve train are comparable.

4.1.2.1 Temperature Measurements

Each temperature measurement valve has eight measuring positions. Four of these measuring positions are similar to the positions described in Sect. 3.3.2 where one sensor is placed in the valve plate bottom midpoint. The other three sensor positions are on the valve seat. All of these sensors are underlying the surface. The spacing between plate surface and plate centre sensor position is 2 mm, whereas the spacing between the seat sensors and the surface is 1.5 mm to ensure a proper valve closing during engine operation.

The further four sensor positions are distributed over the valve stem length. The lower measuring position is in the radius middle between valve stem and valve plate, the upper measuring position is in the middle part of the valve stem which never leaves the valve guidance. The remaining sensor positions are between these upper and lower position, whereas the spacing increases with valve stem height as the temperature changes decrease with increasing distance from the combustion chamber.

The sensors are of type NiCr-Ni thermoelement, their cables run along grooves in the valve stem and are led out on the top of the stem.

The main aim of this setup is the generation of boundary conditions for a subsequent steady-state thermal analysis to generate a temperature field for the valve as an input parameter for the fatigue strength assessment. The temperature of the component determines the material strength, because it decreases with increasing temperature.

4.1.2.2 Strain and Valve Dynamic Measurements

The strain and dynamic measurements include not only the valves, but also further valve train components. The setup expense for valve strain measurements is increased compared to former measurements described in Sect. 3.3. Besides the strain gauges at the valve plate radius, there are two further strain gauges in the valve stem middle. The position of sensors in the stem middle is located in the contact area between valve stem and valve guidance. Therefore, these sensors are countersunk to the stem surface to ensure an even surface in this area. In contrast to former measurements, a modification of the guidance that comes along with a smaller guiding area is avoided. The notches for the strain sensors decrease the stem diameter in this area. For the reason of valve stem stability, two instead of four strain gauges are applied.

As a disadvantage of this setup, it is not possible to calculate circumferential strains in this position, because at least three sensors at different circumferential positions would be necessary to do so. To compensate for this disadvantage, the middle stem strain gauges position points in the assumed main valve bending direction, so that measured strain signals are as close as possible to the real maximum strain in the stem middle. The direction is assumed from the contact marks on the flame ring shown in Sect. 1.3.2. These sensors add additional findings on the valve secondary dynamic. The measured strain shows how much the valve is bended and it also shows whether the assumed bending direction is the real bending direction or not.

Figure 4.6 shows the notch on the valve stem with the applied strain gauge. In a subsequent application step, the notch is filled with epoxy. At the valve plate radius position four strain gauges can be applied. The sensors are not countersunk into the surface, because the valve has no contact to other components in this area. By using four sensors, circumferential strain can be determined from the measurement results. Because of the high temperatures in the area of the valve radius that reach almost 500 °C, regular strain gauges can not be used. Instead, the piezoelectric elements which the strain gauges consist of are embedded in layers of ceramic coatings that protect the sensors and separate the sensor from the hot valve steel surface. Figure 4.7 shows the applied strain sensor in the valve plate radius. For the 49/60 valve, the sensor position is just above the radius because of the conic stem bottom profile. This stem shape causes a large diameter in the radius, therefore the highest valve strains occur above the radius. Hence, this is assumed to be the most fatigue critical area of the valve.

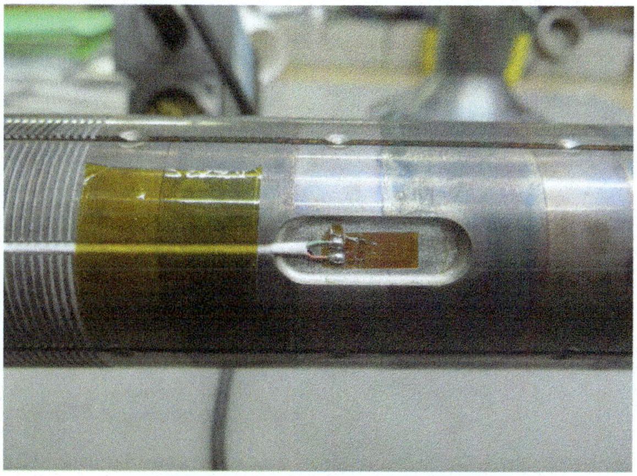

Fig. 4.6 Strain gauge application in valve stem middle

Fig. 4.7 Applied sensor valve plate radius

Both strain measurements in the stem middle and in the valve radius deliver results to compare to simulation results. A good accordance of this comparison verifies the simulation model. A pressure sensor measures the cylinder pressure. The pressure sensor is a standard sensor on the test bed engine.

Moreover, an optical sensor applied in the valve guidance measures the valve lift. The sensor detects small cogs on the valve stem as they move past the sensor during valve lift. The position of the cogs on the valve is shown on the left in Fig. 4.8. The adjacent view shows the geometry of the cogs in detail. Furthermore, regular

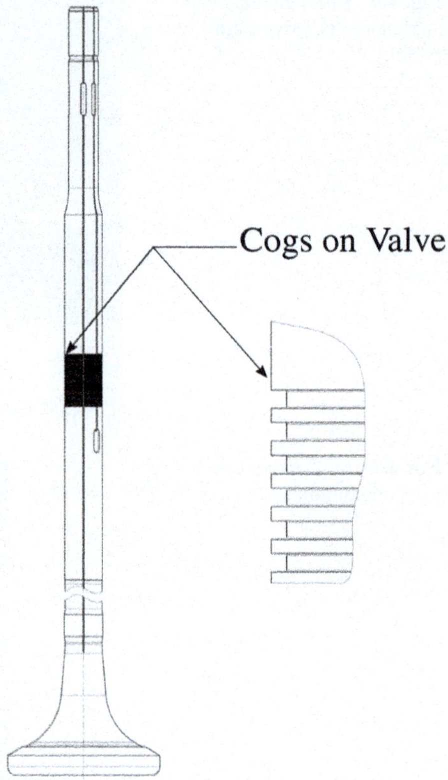

Fig. 4.8 Cogs on valve stem for valve lift measurement

strain gauges are applied to the pushrod and the valve springs to determine the forces transmitted by these components. The temperature here is below 100 °C, the strain can directly be transformed to a force using formerly determined calibration values.

4.1.3 Applicability of Measurement Results

The measurement valve has a slightly different geometry compared to the serial valve design. The main difference between serial and measurement valves are the notches on the measurement valve for the cable duct.

However, the serial valve geometry is applied to the MBS model for two reasons. It has a simpler geometry, which simplifies the implementation to the model and reduces simulation time. Furthermore, the approach aims to evaluate serial valve designs and not measurement valve designs.

Table 4.1 Mode comparison

Elastic mode number	Frequency deviation (%)
1	+0.6
2	+0.2
3	+0.7
4	+0.2
5	+0.8
6	+0.1
7	+0.2
8	+.1
9	+0.5
10	–0.3
11	+1.1
12	+0.4

Nevertheless, the verification of the simulation model utilises the measurement data. Therefore, it has to be ensured that the notches do not influence the deformation results significantly.

Two FE models of a 49/60 exhaust valve, one with the series valve geometry and the other one with the measurement valve geometry, are utilised to compare the results. Firstly, a modal analysis is performed for both models. For similar modal results, the resulting deformation is similar because the deformation of the MBS valve is derived from the modal shapes (see Sect. 2.1.2.2). The results of these analyses are shown in Table 4.1. The resulting modes are similar. The small deviations of approx. 1% are acceptable.

The serial valve is rotational symmetric resulting in pairwise appearing modes, while the measurement valve is rotational slightly asymmetric because of the cable notches. Therefore, the eigenfrequencies do not appear twice and the deviation varies from mode to mode with different deviations.

Moreover, a fictive bending load is applied to the valve plate to compare the resulting strain in the valve plate radius. The difference in the resulting strain is very small, the maximum strains differ by about 0.4%. It is important that this strain value of the measurement valve matches the strain value of the serial valve as this is the value that defines the load for the fatigue strength assessment. The strain sensors are applied to the surface of the valves, therefore there is no impact of notches at the measurement position. The results of the analysis are shown in Fig. 4.9.

The distribution of the resulting stress is very similar in both models. The maximum and minimum stress only differ slightly from each other. Therefore, the load determination using the measured strain signals of the sensors in the valve plate radius delivers values that are comparable to the serial valve geometry. The second position of strain measurements is the stem mid position. These signals are utilized to verify the overall valve dynamic. Therefore, it is important that the general char-

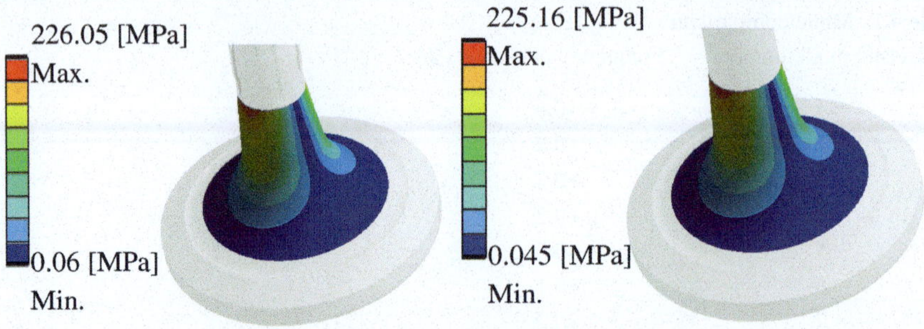

(a) Strain results of valve with measurement geometry (b) Strain results of valve with serial geometry

Fig. 4.9 Comparison of stress in valve plate radius

acteristics of this value matches for measurement and serial valve. Nevertheless, the requirement for an exact match of the strain values is not as high as for the strain of the valve plate radius. Figure 4.6 shows that there is a notch in the stem middle. In reality, the notch is filled with epoxy after the strain sensor application, therefore the notch impact is decreased compared to the case shown here. The epoxy-filling is not considered in the FE model, resulting in a more conservative estimation of the notch influence.

To compare the valve geometries, a fixed boundary condition is applied to the upper part of the valve stem, right above the guidance area of the valve stem. A tension force is applied to the valve plate bottom. Figure 4.10 shows the results of this analysis. The resulting normal stress in the valve is constant for valve sections of the same diameter. This results in a constant stress of approx. 49.2 MPa for the valve stem with serial geometry in the guidance contact area. The model with the measurement geometry shows similar results apart from the oval notches where the strain sensors are applied. In the inner edges of the notches there is a stress concentration. The bottom of the notch shows only slight stress concentration of 6% compared to the serial geometry stress. This is the position the sensors are applied to. As the strain data from these positions is not used for the fatigue strength assessment, the deviation of 6% in this conservative modelling, which does not consider the epoxy-filling of the notches, is acceptable. The resulting error is small enough to use the strain signals for simulation model verification.

The notches of the measurement geometry influence the overall valve stiffness. Therefore, the setup from the previous analysis is used to determine the differences in deformation. With the applied tension force the resulting axial deformation is approx. 1% higher for the measurement geometry. This slight deviation is also acceptable.

The comparability of the temperature distribution of the exhaust valve during engine operation is also compared by an FE analysis. Thus, fictive thermal boundary conditions are applied to the model. A convection is applied to the valve plate bottom. A temperature of 150 °C is applied to the guidance contact area in the middle of the valve stem, while a temperature of 500 °C is applied to a small area on one side of the

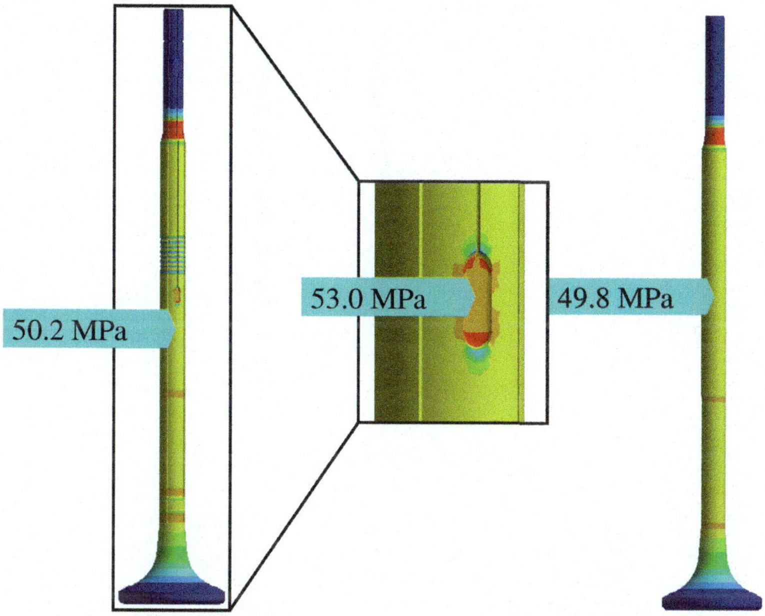

(a) Valve stress of measurement geometry

(b) Valve stress of serial geometry

Fig. 4.10 Comparison of stress in valve stem middle

valve plate side to simulate an asymmetrical temperature impact. Both models show very similar results. There are only small deviations of temperature distribution as Fig. 4.11 shows.

This section showed the investigation of differences between the serial valve geometry and the measurement valve geometry regarding their mechanical and thermal behaviour. For both the mechanical and the thermal loads, the results of the FE analyses showed similar results, there are only small deviations of the resulting strains and temperatures. These deviations are acceptable. In conclusion, the measured results of valve strain and valve temperature can be transferred to the serial geometry model.

4.1.4 Execution

The following section describes the process of measurement execution and explains the different approaches of measurement data recording. The general execution of measurements mainly follows the former approach described in Sect. 3.3. Firstly, the section describes the execution of temperature measurements. These measurements deliver quasi-static temperature data. Secondly, the strain measurement execution

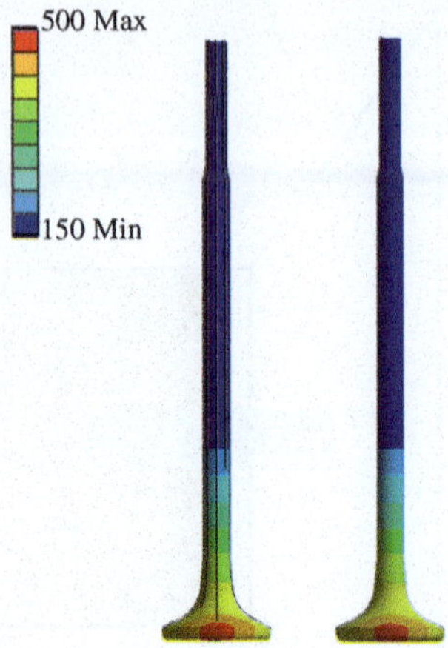

(a) Temperature distribution of measurement valve and serial valve

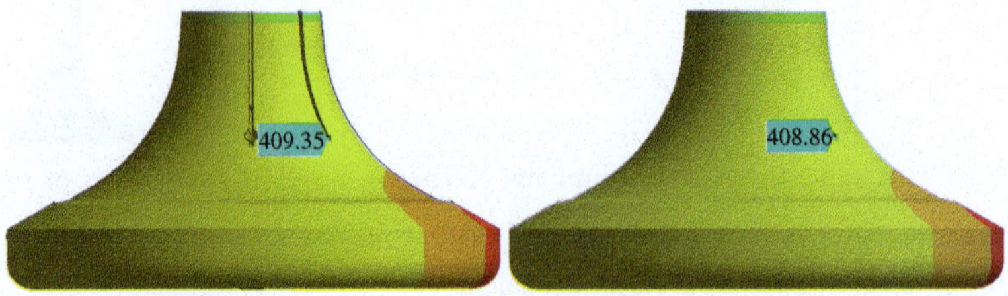

(b) Temperature of measurement point 8, measurement valve geometry

(c) Temperature of measurement point 8, serial valve geometry

Fig. 4.11 Validation of valve measurement geometry

and the recording of data is described. The strain measurements are dynamic. Two types of measurement data are generated, the SSM data and the MSM data.

4.1.4.1 Temperature Measurements

The component temperature changes slowly. For a constant engine operation point, the component temperature converges towards a constant value that shows only small fluctuations after a certain latency. The temperature changes of components are too slow for dynamic measurements, so only static engine operation points are part of the temperature measurement execution. For each of the measured engine operation points (definition see Sect. 4.1.4.2) a latency of 20–30 min is adhered to, to ensure a constant and stabilized temperature for each sensor position. The recording of the temperature values starts after this latency.

4.1.4.2 Strain Measurements

In contrast to the temperature, the movement of the valve train components and thus the resulting strains and forces show a highly dynamic behaviour.

The engine is operated in diesel mode for the measurements. With respect to operating points, the aim is to have a wide base of data, but keep the measurement time short to decrease the risk of failure of strain sensors.

The rotational speed is changed from 600 to 420 rpm in six steps, the load is varied from 110 to 0%. Also idle load points at engine overspeed are included. The measurement of several potential combinations of load and engine speed contributes to the generation of the valve load spectrum. This load collective delivers a verification base for simulative generated load collective.

All engine parameters remain as constant as possible during these measurements. Each dynamic measurement is performed with all measurands, while 20 segments are recorded during the SSM. The chosen number of segments serve as a control that valve dynamics are repeatable at a constant engine operation point. One segment includes four combustion cycles, so that the behaviour at the begin and the end of a combustion cycle are recorded properly. The resolution per channel is approx. 17 kHz.

The second measurement type for each operation point, the MSM, delivers data for the statistic evaluation. Several measurands are switched off. Only the valve lift, valve strain gauges and ignition pressure are active measurement channels which results in an increased resolution of approx. 25 kHz for these channels. The focus of the statistic measurements is the detection of strain maxima during valve closing. Therefore, a segment includes ca. 540° crank angle (CA) of a combustion cycle, whereby the gas exchange phase is recorded completely. The statistical evaluation of the maximum strain requires a higher number of samples. Therefore, the number of segments is increased to 1000 to reach 50 specimen for the GPD, where only the maximum 5% of all strain maxima define the shape of the distribution (see Sect. 4.1.1.2).

4.1.5 Evaluation

This section describes the evaluation of measurements which were described in the previous section. The evaluation of the measurements consists of the three separate parts. These parts are the temperature data evaluation, fatigue data evaluation and dynamic behaviour data evaluation. The aim of the temperature measurement is the generation of a temperature field to consider in the subsequent fatigue strength assessment. The temperature data evaluation uses only the valve temperatures and assumes a static temperature state during the engine operation.

The fatigue data evaluation mainly focuses on the evaluation of maximum strain occurring at the valve during operation. The distribution of maximum valve load determines the load spectrum that has to be considered in the fatigue strength assessment.

Furthermore, the dynamic measurements deliver a data base for the validation of the simulation model of the valve. The dynamic behaviour data evaluation analyses several aspects of the overall valve dynamic including the direction of valve closing, the dominant engine orders in the valve strain spectrum and the correlation of engine operation parameters and valve dynamic.

4.1.5.1 Temperature Data

The valve temperature data from the measurements delivers target values for the calculation of an overall valve temperature field. The mechanics department of MAN ES provides an APDL based script for calculation of temperature fields of the cylinder liner. The liner and the valve have in common that they are rotationally symmetric and have a non-symmetrical temperature distribution. The script is therefore adapted to the valve. The boundary conditions that are adjusted during this adaption are the segmentation of the valve to three temperature zones and the adaption of the angle of hot air contact. The valve is split up to three temperature zones. The valve plate is part of the hottest temperature zone. The plate is much hotter than the rest of the valve because it is in direct contact with the hot air. The valve middle is much cooler and reaches a temperature maximum of ca. 100 °C because this area has no contact with hot air. From the stem middle to the top of the valve the temperature falls below 100 °C. The third part is the valve part from valve radius to valve stem middle where the temperature decreases from the radius towards the stem middle. However, the lower part of the stem shows higher temperatures than the plate top side because it is located directly in the hot exhaust stream. The hot air hits the valve one-sided resulting in an unsymmetrical temperature distribution in the lower part of the valve. The definition of angular temperature boundary condition considers this effect.

Temperature result points on positions equivalent to the real measurement valve sensor positions are applied to the FE model of the valve body in the Ansys *Steady-State Thermal* module. The parameters of the APDL script regarding the symmetricity of the temperature distribution and temperature offsets are iteratively determined by comparison of the measured temperatures at the eight measurement points and the temperature result points.

The resulting temperature distribution including the range of temperatures are shown in Fig. 4.12. It needs to be mentioned that the real temperature distribution of serial engines is probably more symmetrical, due to the valve rotation induced by the rotocap. This can not be verified by measurements because a measurement setup for a rotating valve is impossible due to the cables which are running along the valve

Fig. 4.12 Temperature distribution of the valve

Table 4.2 Simulated temperature valves

Sensor position	Deviation (%)
1	+1.3
2	+0.06
3	+1.1
4	−1.6
5	−0.7
6	−0.7
7	+0.09
8	−0.5

axis. Therefore, the asymmetrical temperature distribution is the more conservative approximation of valve temperatures with a significant hot spot. Table 4.2 shows the percentage deviation of the simulated temperatures from the measured temperatures. The deviations are within the range of measurement inaccuracies.

4.1.5.2 Dynamic Valve Train Behaviour

This section shows the evaluation of the dynamic valve train behaviour from measurements.

At first, the valve clearance during engine operation is determined by utilizing the full valve train model, also referred to as CVT. The clearance can not be measured during engine operation directly. It has to be determined from resulting pushrod forces.

This is one of the few cases in the present thesis, where the CVT is utilised for analysis because the pushrod is needed in the simulation. The comparison of the simulated and measured valve lift verifies this determination. This evaluation delivers the important boundary condition of the real valve clearance. The input is necessary for an accurate simulation model. Furthermore, the dynamic analysis includes the analysis of the valve closing direction. This analysis shows whether or not the valve has a dominant closing direction which can deliver information as to the cause of the valve dynamic. In contrast to the evaluation of the valve clearance, this evaluation delivers no direct boundary condition, but information on the unknown external influences on the valve.

Valve Clearance

The valve clearance is adjusted during cylinder head installation. The inlet valve clearance is 0.5 mm, the exhaust valve clearance is 1.3 mm. The valve clearance changes during engine operation because of temperature dependent deformation of engine parts with respect to each other. Above all, at exhaust side a significant reduction of valve clearance can be observed towards high engine loads. A further change of the valve clearance can appear due to wear of the components. This phenomenon can indeed not be detected during the measurements, because the time period of the measurements is too short to detect wear phenomena.

The comparison of the measured and the simulated pushrod force detects the change of the nominal valve clearance due to engine operation temperature. The pushrod force is used to adjust the clearance of the simulation model to the real measurement clearance. By the comparison of the measured and the simulated pushrod signal the clearance is obtained iteratively. The result of the valve clearance evaluation is shown in Fig. 4.13. The black curve is the measured pushrod force and is therefore the reference. It is clear that the stand still clearance of 1.3 mm results in a too high pushrod force. With a clearance of 0.5 mm the resulting pushrod force shows a good accordance with the measurements.

Valve Lift

After the valve clearance adjustment, the measured and simulated valve lift should be equal. The lift of both valves of the CVT is exactly equal, because no tilting of the rocker arm or valve bridge is considered in the model. Therefore, on the simulation side, only one valve lift is shown. However, in reality there might be a difference between the valve lift curve of both valves. Therefore, both measured valve lift curves are used for the comparison with the simulation model. Figure 4.14 shows slight deviations of the measured valve lifts of exhaust valve 1 and 2 and also

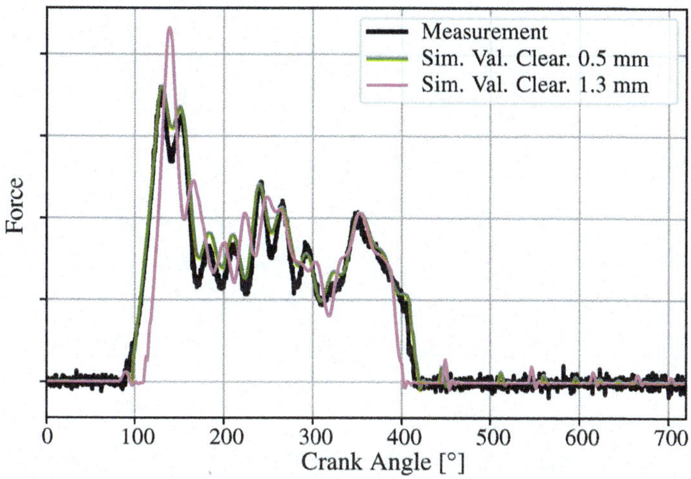

Fig. 4.13 Comparison of pushrod force

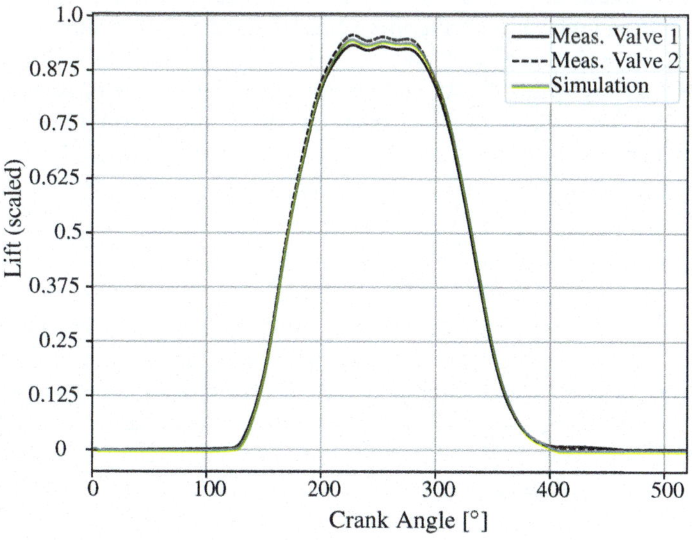

Fig. 4.14 Comparison of valve lift

the simulated valve lift. These deviations are explainable: In reality, a slight valve bridge tilting can occur. The tilting manifests itself in a deviation of the valve lift of the valves actuated by the valve bridge. Therefore, the real valve lift is not identical for both valves but differs approx. 1 mm for the mean valve lift. As the simulated valve lift lies between the measured lift of valve 1 and valve 2, a good accordance of simulation and measurement can be assumed for an ideal valve bridge without

Fig. 4.15 Sensor position valve stem middle

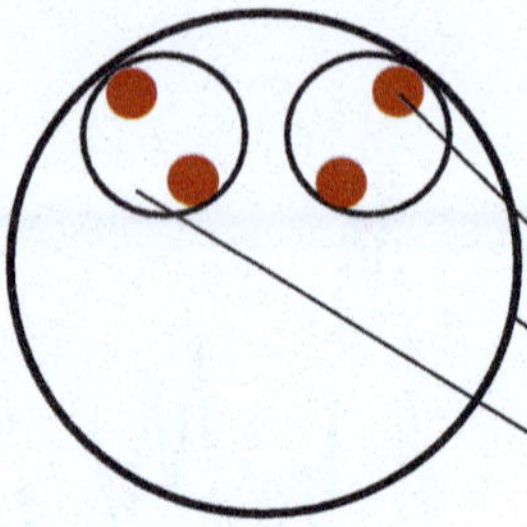

misalignment. The experimentally determined valve clearance is validated by this evaluation and can now be applied to the OVM.

Valve Stem Middle Strain

The strain measurements on the valve stem middle deliver another component of the simulation model validation (Fig. 4.15). The variance of the stem middle strain is small, especially in comparison to the radius strain variance. This shows that the secondary dynamic is relative constant during the engine operation for a constant engine operation point. It is furthermore clear that the sensor plane is not the bending axis, because the signals of both opposite strain sensors do not have an equal amplitude with opposite sign. A bending can indeed be assumed based on the strain characteristics. These findings match the results from Fig. 4.16 that already indicated a dominant bending direction that is not the presumed one. Figure 4.16 also indicated that the bending direction is not constant for different engine operation points.

Valve Closing Direction

The evaluation of the closing direction shows whether the valve closing appears directional or evenly distributed. The strain measurements on the valve plate radius enable an analysis of the strain on every point on the plate radius circumference. The evaluation of the strain data uses the same equations already introduced in Sect. 3.3.3.

The valve strain is evaluated for the small interval of maximum peak strain during valve closing. By using the above mentioned equations, the strain is evaluated in one degree steps, subsequently the maximum strain value and its angle around the valve plate radius are determined. It is therefore assumed that the maximum peak occurs during the first contact of valve and valve seat ring. Consecutively, the direction of the maximum peak strain is the direction of the valve closing.

The results of this analysis show the exhaust valves have slight deviation from the radial cylinder direction that was predicted as the main bending direction from the flame ring contact of the exhaust valves.

There are several conceivable influence factors on the main valve bending direction that include the direction of the dynamic exhaust stream, the temperature induced deformation of the valve guidance and the valve seat ring, transverse forces on the valve caused by valve spring misalignment and the stiffness of the valve bridge.

Figure 4.16 shows the distribution of the first contact of valve and valve seat ring for the exhaust valve 1 for different engine speeds and a constant engine load of 75%. It is clear that some rotational speeds come along with a more directional closing than

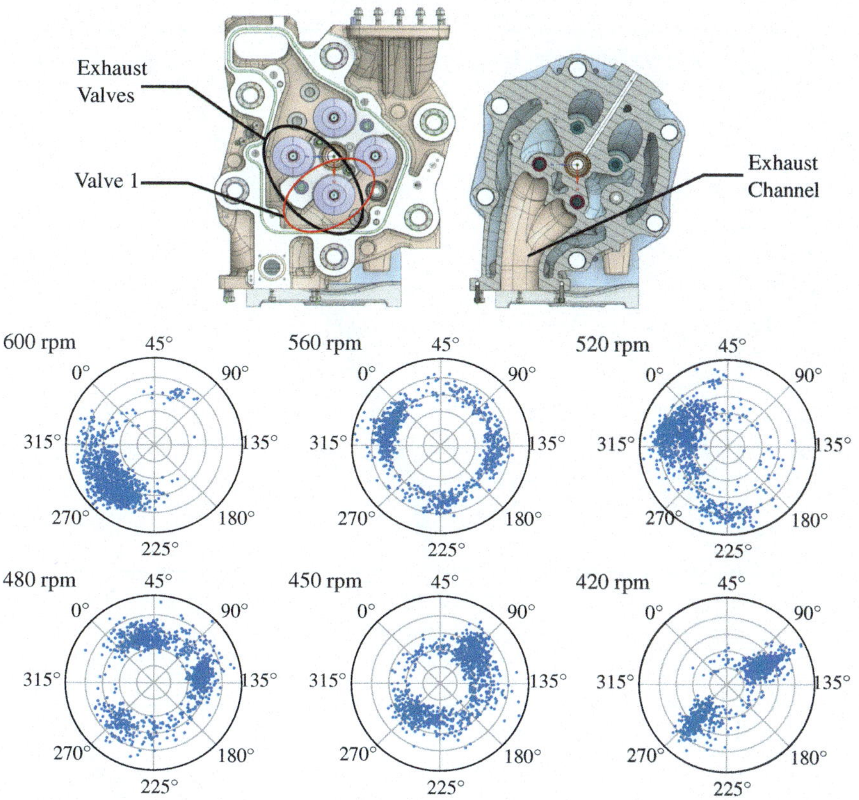

Fig. 4.16 Closing direction of exhaust valve 1

other rotational speeds. Especially 600 and 420 rpm come along with a directional valve closing while 560 rpm shows an almost random direction of valve closing. This leads to the assumption that the valve train excitation is influenced by the rotational speed of the engine. Even lower rotational speeds might therefore result in a highly directional valve closing and potentially to higher valve loads as a result.

Statistic Analysis of The Valve Closing Strain

The 1L 49/60 MSMs have 1000 segments, therefore each GPD based evaluation applying the extreme value distribution uses 50 samples that are the 5% measurement data with the highest strain peaks.

The comparison of the normal distribution and GPD in the strain probability diagram shows that, especially in the tail area, the GPD has a good accordance with the measured strain points in the diagram. On basis of these results, the GPD provides an approach for the assessment of pure simulative generated strain data as an input for fatigue calculation.

Figure 4.17 shows examples of the data fitted extreme value distribution for different engine operation points.

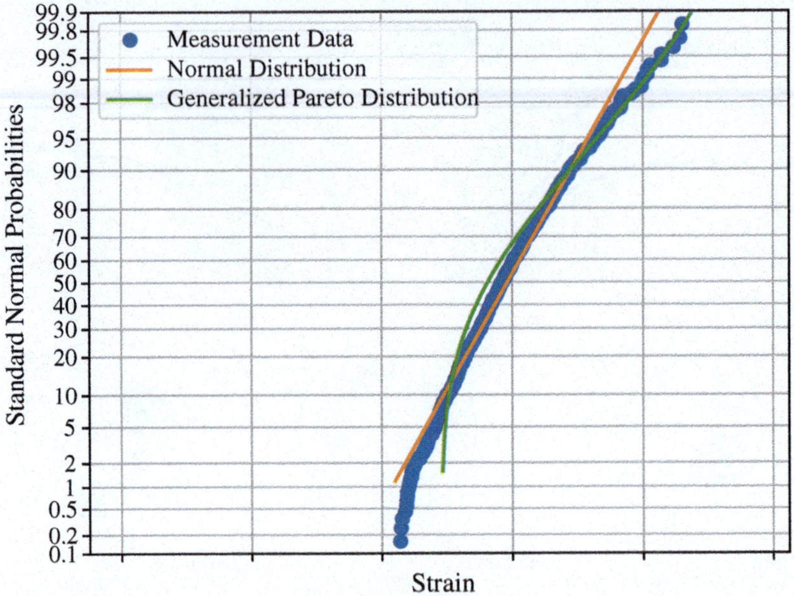

(a) 110% load; 600 rpm

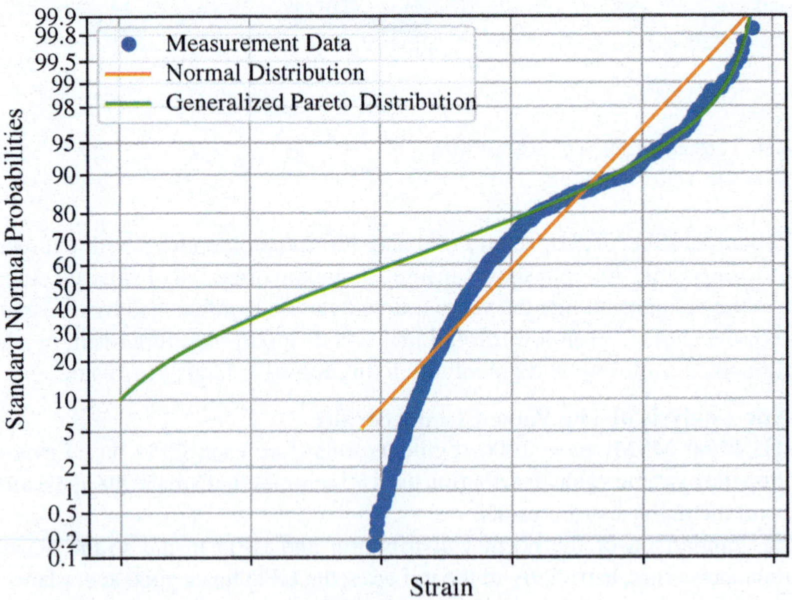

(b) 50% load; 600 rpm

Fig. 4.17 Distribution of strain maxima, 49/60 engine

4.2 Multibody Simulation Model of the Valve

After explaining the measurement set up and the evaluation, this section focuses on the simulation of the valve dynamic. The whole valve train experiences highly dynamic loads during the load exchange cycle. Apart from the primary dynamic due to the valve opening and closing process, experimental findings on heavy-duty engines show a significant secondary dynamic of the load exchange valves. Unlike simulative approaches with the main focus on valve train bearings of the camshaft introduced in Sect. 3.1, the main aim of the simulation model used in this work is to reproduce the dynamic of the load exchange valves as accurately as possible. The valves interact with several other valve train components and are therefore influenced by them. The requirements of the simulation model are the ability to simulate the valve's flexible deformation and to include the relevant contact boundary conditions to all components the valve interacts with. To sum up, the model has to include all relevant physical effects influencing valve dynamic and valve fatigue.

The implementation of a flexible body instead of a rigid body and also the implementation of 3D contact elements lead to increased simulation time. The flexible body increases the number of system DOF and therefore leads to a larger system of DAEs to solve. Furthermore, the flexible body deformation calculation is added to the overall calculation operations. Regarding 3D contact elements, the implementation of EHD requires a coupling of the MBS system and the EHD system, because these systems interact with each other. The necessary Reynolds equations add up the overall number of system equations to be solved and therefore increase the simulation effort, too. A further increase of simulation time is caused by the implementation of a PCM contact. The collision detection and the iterative determination of contact penetration requires several additional iteration steps.

To handle an increasing total simulation time, the number of components in the simulation model is reduced. For this purpose, an OVM is extracted from the CVT. This reduces the system DOF and eliminates contact boundary conditions. Moreover, it is taken advantage of the symmetry of the system since only one instead of both valves of a valve train are implemented. The aim of this reduction resulting in the OVM is to keep the simulation time as short as possible.

In contrast to the OVM approach presented here, the publications focusing on valve and valve train dynamic include the full valve train from cam contour or even camshaft to the valve. The reason for this is that valve train components contact and elastic deformation influence the valve lift and its derivatives.

In this thesis, the following steps were taken to enable the usage of the OVM: Firstly, the stability of the valve lift for a static engine operation point was ensured. The evaluation of measurement data shows congruence of the valve lift curves for all measurement curves. The variance of the valve lift is very small and therefore negligible. Because of the small variance of the valve lift, a single calibration of the OVM is sufficient. Therefore, the verification of the valve lift and the valve clearance with the CVT is performed once according to the procedure presented in Sect. 4.1.5.2. The resulting boundary conditions are applied to the OVM. The CVT does not contain time-consuming modelling elements e.g. EHD contact elements.

4.2.1 Structure of the One Valve Model

The OVM consists of the valve body, the valve guidance, the valve springs and the valve seat. It is extracted from the CVT including all valve train components. The 48/60 OVM extracted from the corresponding CVT is shown in Fig. 4.18. The interface of the OVM and the CVT is the actuation of the valve by the rocker arm or valve bridge, respectively, with the elephant foot. In the OVM the elephant foot is replaced by a dummy body. The positional information of the valve lift on the side of the elephant foot and its derivatives are directly applied to the dummy body's joint as boundary conditions. The dummy body transmits the positional information due to a contact force element to the valve. The resulting force depends on the valve lift and its derivatives that are transmitted to the valve but also on the contact parameters between the dummy body and the valve. Moreover, the OVM considers the stiffness and damping of the rocker arm also influencing the valve actuation force. The OVM includes an implementation of a further dummy body that is connected to the other dummy body and to the valve via additional contact force elements. This modelling of the contact between elephant foot and valve top is explained in more detail in Sect. 4.2.2 and shown in Fig. 4.25 in this section. To ensure the accurate contact force on the valve top, the OVM is verified using the CVT.

4.2.1.1 Valve Body Implementation

There are two different implementations of the valve body. One implementation is used in the 48/60 model, the other implementation is used by the 49/60 model. The difference between the models is small and affects the contact of valve and valve seat ring. The implementation used in the 48/60 uses simple contact elements

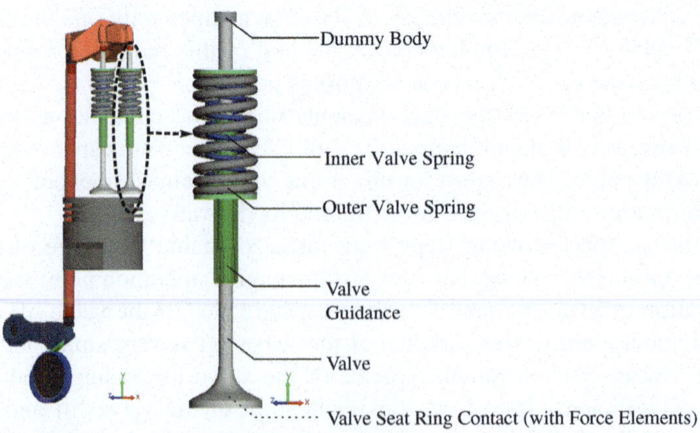

Fig. 4.18 One valve model, valve train of 48/60 engine

and is therefore much more time-efficient. The modelling depth is sufficient for investigations regarding the general valve dynamic. The other implementation uses a 3D contact model and thus takes more simulation time. The additional information is needed for the investigation of the valve load data. The higher resolution of the contact enables a more detailed investigation of the valve dynamic. The modellings are outlined in more detail later in this section. At first, the general valve body implementation is explained:

The valve body is implemented as an elastic-flexible structure to consider the valve bending deformation and extract resulting modal stresses from the valve deformation. The flexible structure is based on a FE model of the valve that is reduced by the procedure introduced in Sect. 2.1.2.2. The referred section shows that it is essential to define a master node for each point a force may be introduced to the valve structure. The master nodes of the FE valve model structure are also used during the reduction process to define boundary conditions of the CMS.

The slave nodes belonging to a specific master node are defined via surface imprints on the body contour in the FE model. All nodes within the imprinted surface are defined as slave nodes. The master nodes are in the centre of these imprints. The connection between master and slave nodes is a flexible connection to enable the flexible structure deformation in the areas of force introduction. The forces applied on the master nodes in the MBS model are distributed to all corresponding slave nodes. This distribution avoids singularities when forces are applied to the master node.

In the following, the differences between the valve body implementations used in this thesis are explained: In the 48/60 model, the valve seat contact area is discretised and twelve master nodes are applied around the valve seat circumference. These master nodes are used to introduce force elements between valve and valve seat ring. The force elements use root functions and therefore allow the discrete contact states slip, stick and a contact lift. This contact formulation is efficient, but has the risk that the contact can not be detected when the valve rotates around its main axis. This contact implementation is place-bound, therefore the valve rotation must be locked. Figure 4.19 shows the distribution of the master nodes on the valve seat contact area. Further details on this implementation can be found in Sect. 4.2.1.3.

For the 49/60 model, the contact between valve seat and valve seat ring is applied via PCM (see Sect. 2.1.4.1). Thus, there is one master node defined for the application of the resulting force due to the PCM. This contact formulation provides the flexibility to rotate the valve relative to the valve seat ring because of the included collision detection algorithm.

Additional master nodes along the valve stem enable the introduction of further forces and interactions. Besides the master nodes used for the seat ring contact of the 48/60 model, seven further master nodes are spread along the stem to introduce interaction forces of:

1. Valve—Dummy Body (Elephant Foot)
2. Valve—Valve Spring Top Plate
3. Valve—Valve Guidance Top

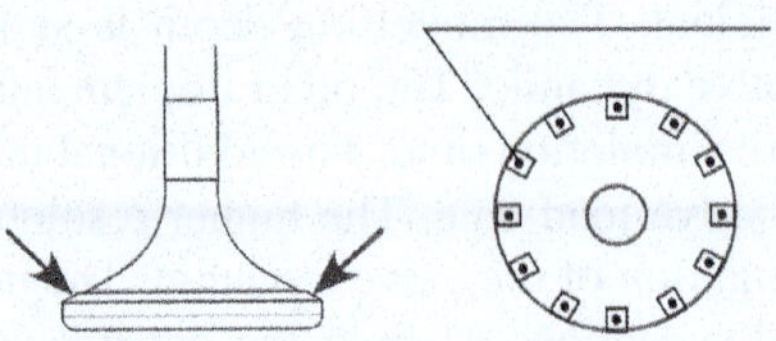

Fig. 4.19 Master node configuration of 48/60 valve seat contact

4. Valve—Valve Guidance Mid
5. Valve—Valve Guidance Bottom
6. Valve—Potential Transversal Valve Bottom Force
7. Valve—Cylinder Pressure.

The implementation of the 49/60 valve includes many more master nodes for the introduction of dynamic fluid forces. Further details on the implementation of additional forces on the valve are described in Sect. 4.3.5.1.

4.2.1.2 Valve Guidance Implementation

To achieve a proper sealing of the combustion chamber and an even distribution of forces during valve closing, the valve plate and the valve seat ring should be concentric. Therefore, it is the guidance's main function to ensure the concentric valve movement. Also the guidance enables the relative movement between the static cylinder head and the valve during valve lift. The guidance has no own lubrication supply. Instead, lubrication oil that is supplied to the top of the valve train drops down the valve and thus reaches the gap between valve and guidance. For this reason, neither the exact amount of lubrication oil in the contact area between valve and guidance nor the thickness of the lubrication film can be determined, only estimated.

The valve guidance is installed in the cylinder head by press fit in the middle part and in the lower part of the guidance which results in a very high stiffness of the guidance and in negligible elastic deformations of the guidance. The guidance thus is implemented as a rigid body in the OVM with zero degrees of freedom. The guidance has a clearance which permits small radial valve movement. The forces and deformations resulting from the contact of valve and guidance have an impact on the valve dynamic during the full working cycle. Thus, an accurate implementation of the effects of oil pressure, valve deformation and resulting contact forces is important.

An EHD contact element meets these requirements. The theoretical background on EHD simulation is introduced in Sect. 2.1.4.2. Two assumptions are made for the lubrication state of the contact: A thin oil film exists but is thin enough, so that the filling level is zero as there is no oil column in the guidance. The oil film thickness is an assumption based on empirical analyses of EHD parameters for components without an own oil supply. The rough surface contact uses the Greenwood-Tripp model. This modelling applies to the case of the valve passing through the oil film, so that valve and guidance surface get in direct contact.

During engine operation the exhaust side of the cylinder head deforms due to the flow of hot exhaust gases. This cylinder head deformation causes deformations of the exhaust valve guidance. The guidance gets a characteristic bended form along its centre line. In addition, the cylindrical cross-section deforms into an oval form. The ovalisation increases with decreasing distance to the combustion chamber. The modal decomposition of these deformations shows the ratios of each deformation mode on the total guidance deformation. Figure 4.20 shows the 3D reconstruction of the first four modal flexible guidance deformations. The first two modes have a relevant influence on the contact conditions as the deformation is in range of the guidance clearance that is depending on the valve train type approx. 0.1–0.2 mm. The deformation amplitude decreases for higher modes. The third and fourth mode may have a minor influence. Because of the small magnitude no influence on the valve dynamic for all higher guidance deformation modes is expected.

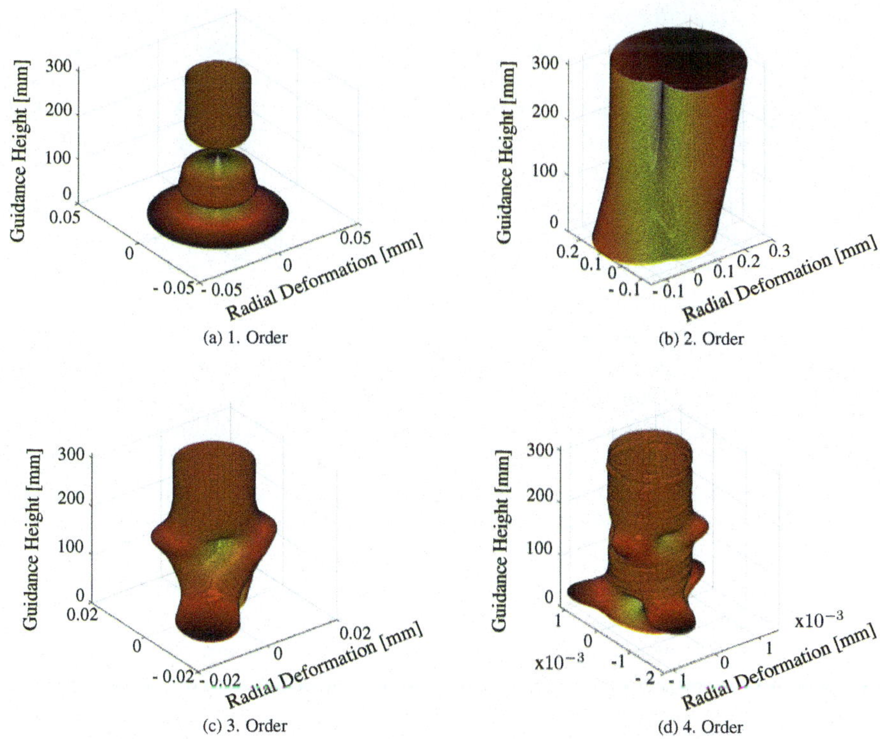

Fig. 4.20 3D reconstruction of guidance deformation modes

The EHD contact has a direct option of including deformations of the contacts shell and journal geometry. The basic geometry is the ideal cylindrical shape of the guidance. By the superposition of deformation functions each mode is added to the ideal geometry individually.

4.2.1.3 Valve Seat Ring Implementation

The valve closing is proven to be the phase causing the highest valve loads. Even in the case of an ideal concentric valve closing, the impulse induces high normal stresses in the valve. In reality, the valve seat and valve seat ring are not concentric for multiple causes. One of these causes is the guidance deformation explained in Sect. 4.2.1.2. Moreover, a non-uniform cylinder head heat transfer occurs during engine operation, resulting in a cylinder head deformation, especially on the exhaust side. The non-uniform heat transfer and the cylinder head deformation lead to a deformation of the valve seat ring on the exhaust side resulting in an angular misalignment of the valve seat ring. This mechanism is pointed out by [3] as a cause for increased valve wear. However, the misalignment also increases the valve loads during closing. The thermomechanical model of the 48/60 cylinder head generated by the Mechanics & Validation department of MAN Energy Solutions provides the resulting seat ring deformation. The valve seat ring model has to be able to assign such a deformation. In addition, the contact between the valve and the valve seat ring has to be able to display the three contact states stick, slip and lift as these appear during each work cycle. There are two options to implement this kind of contact behaviour in SIMPACK. The first option does not need the valve seat ring geometry or even a valve seat body. For this contact formulation the 2D SIMPACK force element named *195* is used. This force element describes the contact behaviour between two surfaces with an included stick-slip model. On the valve side the contact patch is defined by the master node on the valve plate top as already shown in Sect. 4.2.1.1 for the valve side. The corresponding contact markers on the valve seat ring side are markers oriented parallel to the initial coordinate system. Each master node on the valve seat has a corresponding contact marker on the valve seat ring side. For the consideration of variable angular misalignment of the valve seat ring it is possible to parametrise the angular orientation of these initial coordinate systems markers so that a specific valve seat ring angle can be assigned.

The second option for the valve seat ring implementation uses the valve seat ring geometry, therefore the valve seat ring is implemented as rigid body in the OVM with a zero DOF joint. For the contact formulation a PCM is applied. Section 2.1.4.1 explains the theoretical principles of this contact formulation. The 3D-geometry of the contact partners defines the contact surface. Markers and master nodes in the contact region are not necessary for this implementation. As the contact detection occurs in the whole contact area it is not necessary to fix a valve DOF.

4.2.1.4 Valve Spring Implementation

The explanations in Sect. 3.1 already show that there is no standardized approach regarding valve spring implementation. The modelling depth has a high influence on the calculation effort. However, the valve spring dynamic influences the valve dynamic significantly. Thus, the implementation for this model should consider all relevant dynamic spring effects. A relevant effect is the local coil contact that influences the spring characteristics. During the valve lifting phase not only the lower and upper coils can come in contact. The middle coils might also come in contact as the local modes appear during work cycle and emphasise vibrations of certain coils. The coils in contact change the overall spring stiffness hence the implementation needs enough contact force elements distributed over each coil.

Several valve spring implementations use an uni-axial valve spring model with infinite stiffness in transverse direction and leave the three-dimensional spring dynamic unconsidered. An example for this implementation is the spring used in [4]. However, modal analysis of the springs indicate transverse modal vibrations in frequency ranges relevant for engine operation.

In addition, springs have relatively wide manufacturing tolerances that affect the valve spring straightness deviation. A spring tilting leads to a transmission of transverse forces from spring to valve. The scheme of this mechanism is shown in Fig. 4.21. Within the scope of valve secondary dynamic analysis the magnitude and overall influence of this transmitted force is analysed. In conclusion, a spring implementation that meets the requirements of an accurate simulation of valve dynamic consists of a linear-elastic body with transverse and axial stiffness and discretised coil-coil contacts.

Using SIMPACK, there are two options of implementing such a spring model. The first method is the SIMPACK-internal Simbeam flexible spring (SFS). The coils consist of several beam elements and contain contact force elements for the coil-coil contact. For an accurate reproduction of the spring stiffness it is important that enough beam elements per coil are used. The implementation process is simple as SIMPACK contains a generator for several spring implementations.

The other option is the spring implementation as a regular flexible body similar to the valve body. This implementation is work-intensive as the geometry and all master nodes for the coil-coil contact elements have to be created in a FE-software first. The accuracy of stiffness in this context depends on the mesh discretisation.

Both implementations have contact elements between the upper coil and the upper spring plate respectively the lower coil and the lower spring plate. Similar to the coil-coil contact elements, the contact elements' working direction is the main spring axis. There are further force elements with a radial working direction. These radial contact elements are only in the area of the top two to three coils respectively the lower two to three coils. These elements ensure that the position of the spring relative to the spring plate stays constant during the work cycle. They limit the transverse spring movement in these areas. It is possible to set a clearance to this radial contact to consider the spring tilting caused by manufacturing. This implementation leads to a transmission of transverse forces to the valve as it is expected in reality. The upper valve spring

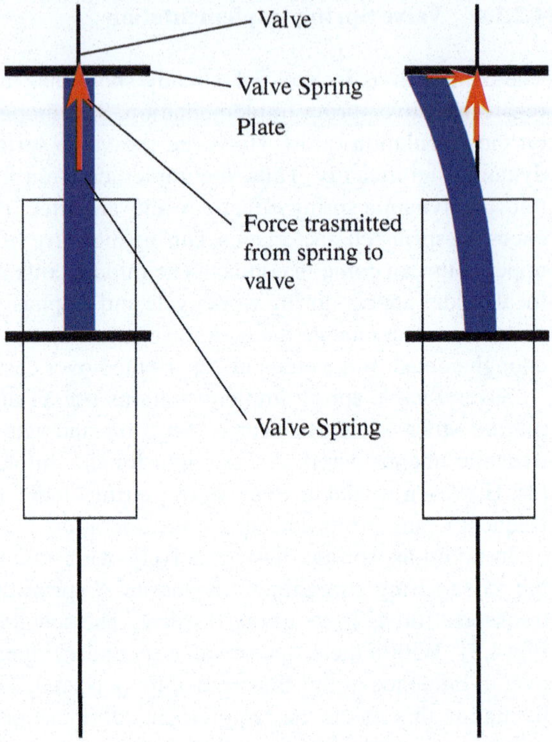

Fig. 4.21 Force transmission from spring to valve

plate is connected to the valve. The corresponding valve marker is defined on a valve master node. The connecting joint is a zero DOF joint. This connection replaces the cone used in reality for the rigid connection of valve and spring plate. The bottom spring plate is connected to the guidance by a zero DOF joint.

Both spring models, the SFS and the flexible body spring, are validated by the comparison of the model with the spring data sheet. The values of interest in this context are the pre-load force F_1 and its corresponding spring length l_1, the maximum force F_2 and its corresponding length l_2 and the spring frequency spectrum. Figure 4.22 shows the results of the spring verification for the SFS as well as for the linear-elastic flexible spring (LEFS). The red curves show the reference values for force and frequency given by the spring data sheet. Regarding the spring force, both spring implementations have a very good accordance with the data sheet values. Both spring implementations show similar frequency characteristics. They differ in frequency only for the spring compression in the range from 55 to 60 mm. Both spring frequency characteristics show slight deviations from the data sheet values. There are two possible reasons: The first reason is the discrete application of the test excitation which leads to discrete frequency levels in the evaluation in Fig. 4.22. A higher resolution of excitation levels in the range of l_1 and l_2 may therefore lead to a better accordance of the simulated and the real frequencies. A second possible

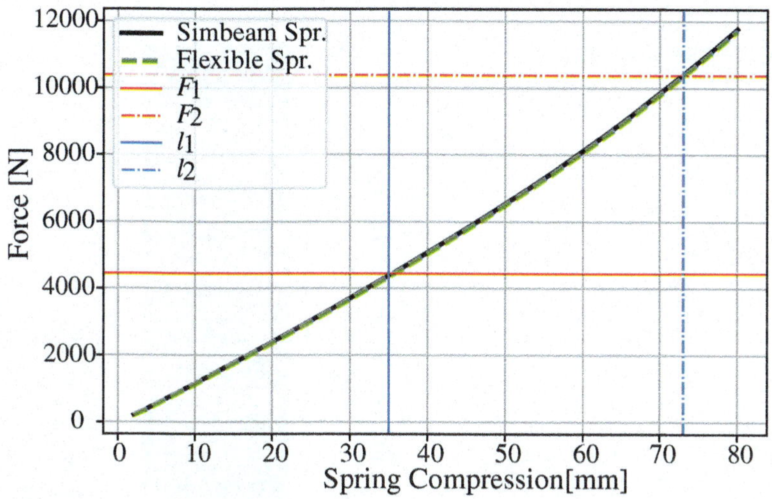

(a) Force characteristics over spring compression

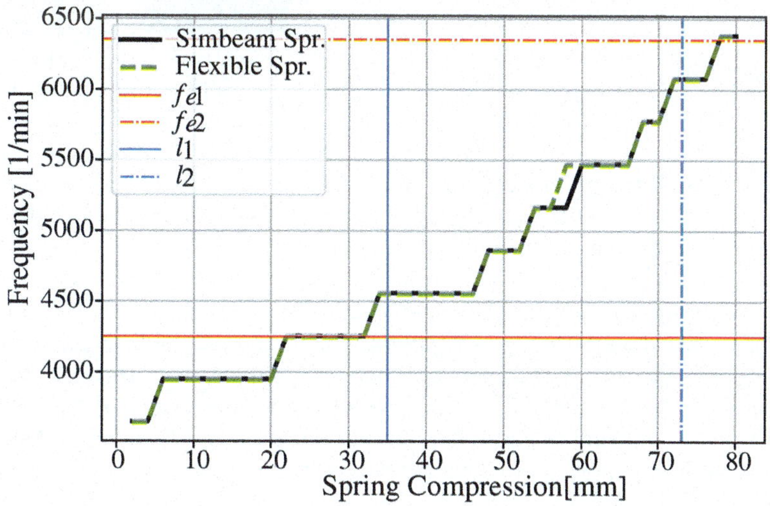

(b) Frequency characteristics over spring compression

Fig. 4.22 Force and frequency validation of spring modellings

Fig. 4.23 Comparison of simulation time with spring model variations

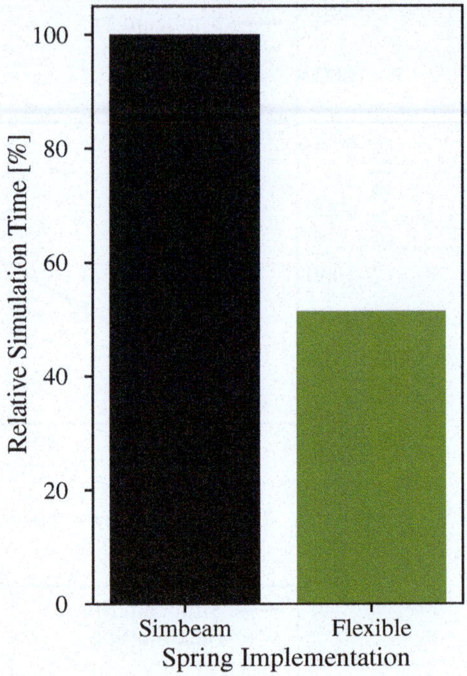

reason is that inaccurate assumptions were made on damping effects. The damping influences the response frequencies of a structure. The real damping of the spring structure is not known. In the model the damping is based on standard assumptions and values of steel damping. The structural and real damping are transformed to an equivalent modal damping. The conversion of structural to an equivalent viscous damping is based on following assumption: The viscous damping and the structural damping are equal when the excitation frequency equals the resonance frequency of the system or component. Deviations from this assumption lead to errors in the conversion of structural to equivalent viscous damping and therefore influence the response of the component. To compare the spring characteristics for the valve train application, both spring models are implemented in the OVM separately and the resulting spring forces during the working cycle are compared to each other. These results are shown in Fig. 4.24. The forces of the SFS and the LEFS are similar.

The investigations show that the LEFS provides a good approximation of the real spring characteristics and has the same accuracy as the SFS.

The LEFS has the advantage of higher efficiency. The comparison of the OVM calculation time with either the SFS or the LEFS implementation shows that the simulation time can be halved (see Fig. 4.23). The setup of the OVM is modified for this efficiency comparison: Time-consuming contact formulations like the EHD and the PCM elements are replaced by simple spring-damper elements. This set up ensures that the flexible spring is the element that dominantly influences simulation time.

Fig. 4.24 Comparison of the spring implementations

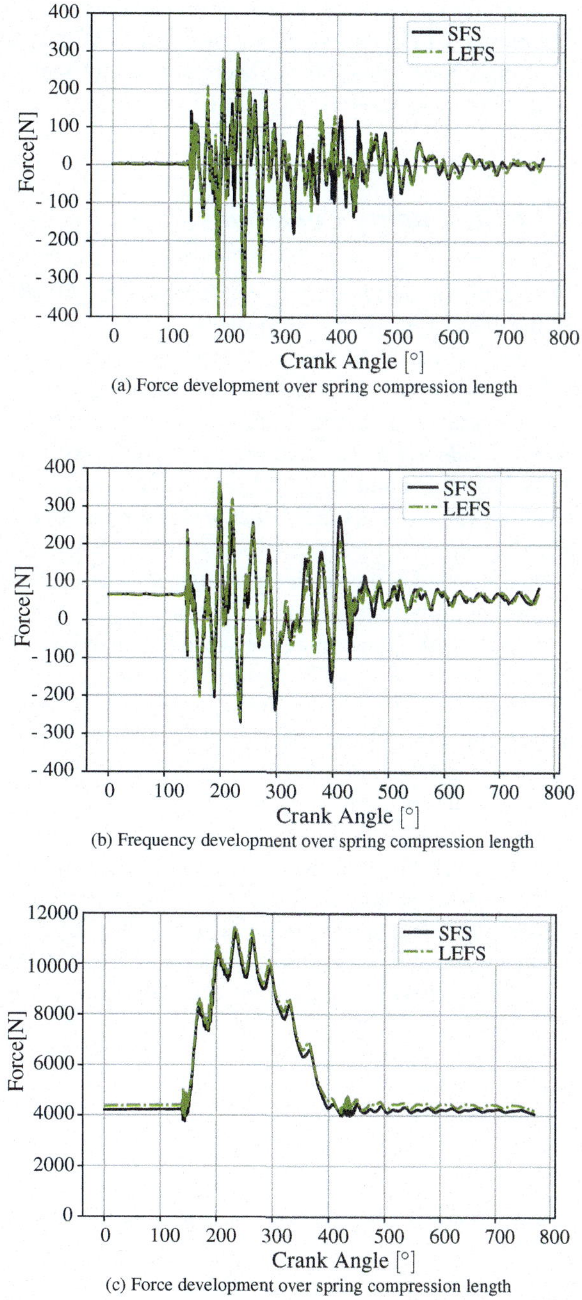

(a) Force development over spring compression length

(b) Frequency development over spring compression length

(c) Force development over spring compression length

The LEFS saves approx. 50% simulation time compared to the SFS in the simplified model. The spring does not interfere with the guidance modelling or the valve seat ring contact modelling. Therefore, additional simulation time for 3D-contact implementations of these elements linearly adds up to the overall simulation time.

4.2.2 *Boundary Conditions*

This section shows boundary conditions of the OVM. Moreover, external influences on the valve dynamic are introduced.

Because the OVM has its system border towards the valve train at the contact point of rocker arm or valve bridge, respectively, the transmitted contact force builds an elementary boundary condition for the OVM. The contact force influences the normal loads in the valve during actuation as well as its lift and lift derivatives. The rocker arm has a non-negligible elastic deformation during actuation influencing the resulting valve lift. As the rocker arm is not part of the OVM, the elastic deformation has to be replaced by accessible modelling elements. The elasticity of the rocker arm is thus included in the valve top contact. Figure 4.25 shows the scheme of this contact. A dummy body transmits the lift information of position, velocity and acceleration to the valve top. The contact between valve and dummy body is applied by a spring-damper force element. Furthermore, a friction force element is applied between these bodies to consider the force transmission between the bodies for a transverse movement of the bodies. The stiffness of contact force elements in general is relatively high. This limits the mutual penetration of the contact partners, because a decreasing of the contact force element stiffness leads to an increasing penetration of the contact partners. If the stiffness is too low, the contact can not be detected. However, the component stiffness of the rocker arm is much lower.

To consider a more resilient valve contact partner, a second dummy body is implemented [6]. This dummy body has its position on top of the first dummy body. Another contact force element is implemented between the dummy bodies. Each contact element has a stiffness that is high enough to ensure a sufficient contact detection with

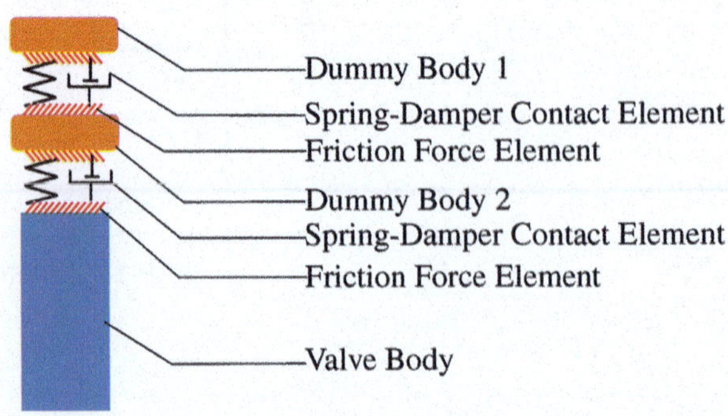

Fig. 4.25 Schematic valve top contact modelling

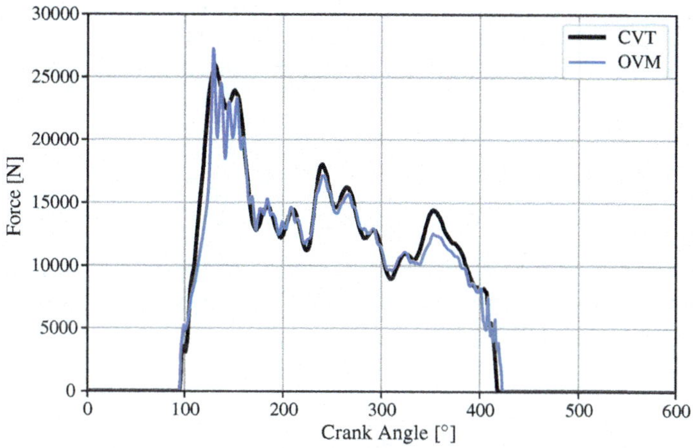

Fig. 4.26 Comparison of contact force on CVT and OVM

small contact penetration. Nevertheless, the overall stiffness of the force transmission decreases as the force elements are in a series connection so that the resulting reciprocal spring stiffness is the sum of the contact force element's reciprocal stiffness values:

$$\frac{1}{c_{tot}} = \frac{1}{c_{cont,1}} + \frac{1}{c_{cont,2}} \tag{4.2.1}$$

To evaluate the accuracy of contact force transmission the contact force transmitted from the elephant foot to the valve in the CVT is compared to the contact force transmitted from the dummy body to the valve in the OVM. From that comparison, shown in Fig. 4.26, a good accordance can be obtained. The accordance of the force peak at the beginning of the valve actuation and a matching frequency in the force progression is good. There is a slight deviation at the beginning of the valve actuation regarding the force frequency that is higher at the OVM. This vibration is negligible because no effect on the valve lift or its derivatives can be observed.

All loads of the valve train that are supposed to be implemented in the OVM must be converted to force components and be applied to master nodes on the valve. For the valve this applies primarily to pressure-related forces. The ignition pressure is applied to the valve plate bottom master node. This force is distributed to all the corresponding slave nodes. All master nodes are in the center of the valve cross-section. The slave nodes are arranged circularly around the corresponding master node. Figure 4.27 shows a valve stem cross section with master node and slave node area. Beside the ignition pressure there are further flow forces acting on the valves. As the exhaust valve opens against up to 20 bar remaining cylinder pressure there is a high axial flow force caused by the exhaust stream as soon as the valve opens. Furthermore, a pressure gradient exists between the side of the valve facing the liner

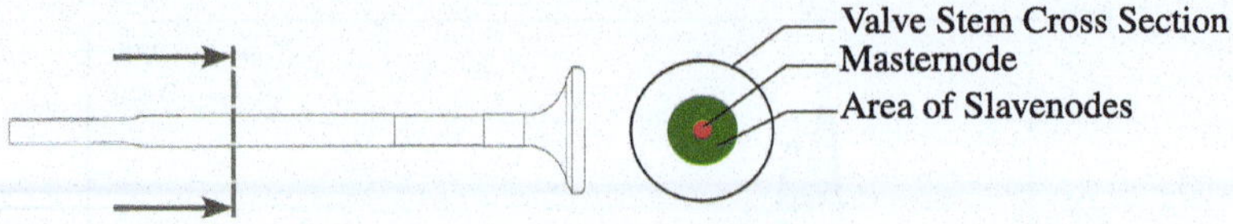

Fig. 4.27 Masternode and area of slave nodes on valve stem cross section

and the side of the valve facing the cylinder centre. This pressure gradient results in transverse forces on the valve. The effect of the transverse flow forces is discussed in Sect. 4.3.3.

4.3 Analysis of Valve Secondary Dynamic Causes

After previous sections introduced the structure of the OVM and its boundary conditions, the resulting model is now utilised to investigate the valve secondary dynamic. Therefore, this section summarizes the various hypotheses on the cause of the valve secondary dynamic. Moreover, the section describes the analyses performed to verify or respectively falsify these hypotheses.

4.3.1 Preliminary Investigation of Valve Plate Deflection

The examination of the flame ring shown in Fig. 1.6 points out that several engine types show contact of the exhaust valve plate and the flame ring. Before investigating the dynamic of valves hitting the flame ring, a static preliminary analysis of this contact event gives an overview over the forces acting on the valve for the static case. The analysis presented here evaluates the static force necessary to bend a valve sufficiently for it to make contact with the flame ring. The analysis utilizes the FE method, performed in the Ansys Workbench module *Static Structural*. The analysis is performed for the 48/60 exhaust valve. Depending on the valve plate and flame ring design version the distance between both of the components varies in a small range. For the following analysis a distance of 4 mm between the outer edge of the valve plate and the flame ring is assumed.

The applied boundary conditions are shown in Fig. 4.28. The valve is assumed to be at its maximum lift position, the guidance area has a fixed support boundary condition. The assumed displacement is applied to the valve plate. The resulting force reaction at the valve plate is approx. 4400 N.

In conclusion, a very high transverse force is necessary for the static case to result in an equivalent deformation observed at exhaust valves of several engines. The appearance of such a high force in the valve train is inconceivable. There is no valve train component transmitting such a high force.

Fig. 4.28 Preliminary investigation on valve plate deflection

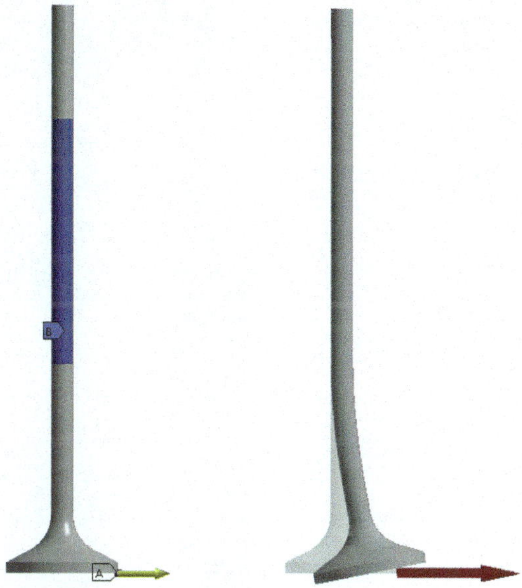

(a) Boundary conditions

(b) Deformed valve

From this finding it can be derived that a static force on the valve can not be the cause for the valve bending. Therefore, the focus of the subsequent analyses is the dynamic of the valve and the valve train. The following sections investigate this dynamic. Similar to the static analysis, the deflection of the valve is one of the criteria to detect potential causes of valve bending.

4.3.2 Geometric Influences

This section describes the influence of geometric deviations on the valve dynamic and investigates its potential of being the main cause of the valve bending vibration. The geometric deviations are caused by either tolerances defined by the construction or deformation caused by engine operation. The expected geometric deviations and changes during engine operation are introduced component-wise.

4.3.2.1 Valve Spring Deformation

The tolerances of the valve springs allow an axial misalignment of the valve spring of up to 6 mm for 45/60 and 48/60 valve springs. The axial misalignment is shown in

Fig. 4.29 Axial misalignment of inner valve spring, 45/60

Fig. 4.29. The upper spring plate is mounted to the valve stem by a cone connector. A ledge on the plate limits the movement of the spring on the plate but leaves a clearance of a few mm. The only force acting between the spring plates and the spring is the friction force. With the plate ledge as a border, the spring can move on the plate. The result of this movement and of manufacturing tolerances is transverse misalignment of the spring regarding the valve axis. Therefore, the spring transmits not only axial forces, but also transverse ones to the valve.

In contrast to the seat ring misalignment discussed in the following section, the valve spring misalignment has no systematic direction. The direction depends on the spring mounting position. Furthermore, transverse force directions of springs add up if two springs per valve are applied. The 48/60 exhaust valves have an inner and an outer valve spring each. However, the springs have opposite winding directions, so that the risk of a positive superposition of forces is minimized.

The fact that the contact marks on the flame ring exhaust side appear on every cylinder of the affected engine indicates a systematic cause with a predominant direction. If the spring misalignment was the cause of the valve bending vibration, the effect would not appear on every exhaust valve of a whole engine at utmost probability, because the direction of the spring misalignment is random. Therefore, the geometrical spring misalignment is excluded from the possible cause of the valve secondary dynamic.

4.3.2.2 Valve Seat Ring Misalignment

As mentioned in the previous section, the valve seat misalignment is a systematic effect and appears on every cylinder. Therefore, this geometric misalignment is a possible cause of valve vibration. The valve seat ring shows mainly an angular misalignment during engine operation. As explained in Sect. 4.2.1.3, the main cause for this angular misalignment is the thermal deformation of the cylinder head. This is the reason for the systematic appearance of seat ring misalignment.

The misalignment has two effects on the valve. During valve closing, the seat misalignment increases the eccentricity of the valve and valve seat ring and therefore increases the bending in the valve radius during closing.

The hypothesis on the dynamic effect of the misalignment applies to the phase post closing: The spring force and the cylinder pressure press the valve in its valve seat ring. With a misaligned seat ring, the valve is bended while pressed in its seat. This increases the valve's potential energy of elastic deformation. When the valve opens, this energy is released and induces the bending vibration.

The determination of the real seat ring misalignment by measurements is not possible. Instead, the misalignment is estimated on basis of experience and FE analysis of the thermal induced cylinder head deformation shown in Fig. 4.30. The thermal analysis of the 48/60 CR cylinder head is part of the design process in the MAN Mechanics & Validation department and is therefore accessible for the present thesis. The analysis shows an angular misalignment of approx. 0.1°. To consider potential deviation of simulation and measurement, the MBS based analysis of the valve seat angle influence considers an angular misalignment up to 0.5 °. The results in Fig. 4.31 show that even for the maximum seat angular misalignment the maximum deflection of the valve plate is under 3 mm. Therefore, the valve angular misalignment is excluded from the possible causes of the valve secondary dynamic. The geometric misalignment induces forces to the valve. In conclusion, their influence on the valve dynamic has to be considered in the simulation model. However, no geometric misalignment is able to cause the observed valve vibrations. Therefore, the next section focuses on external forces acting on the valve.

4.3.3 Quasi-static Flow Forces on Exhaust Valves

The pressure impact of the ignition pressure on the valve is implemented as a force acting in axial direction on the valve plate bottom. However, the identification of a transverse flow force component in Sect. 4.2.2 leads to the hypothesis of a valve

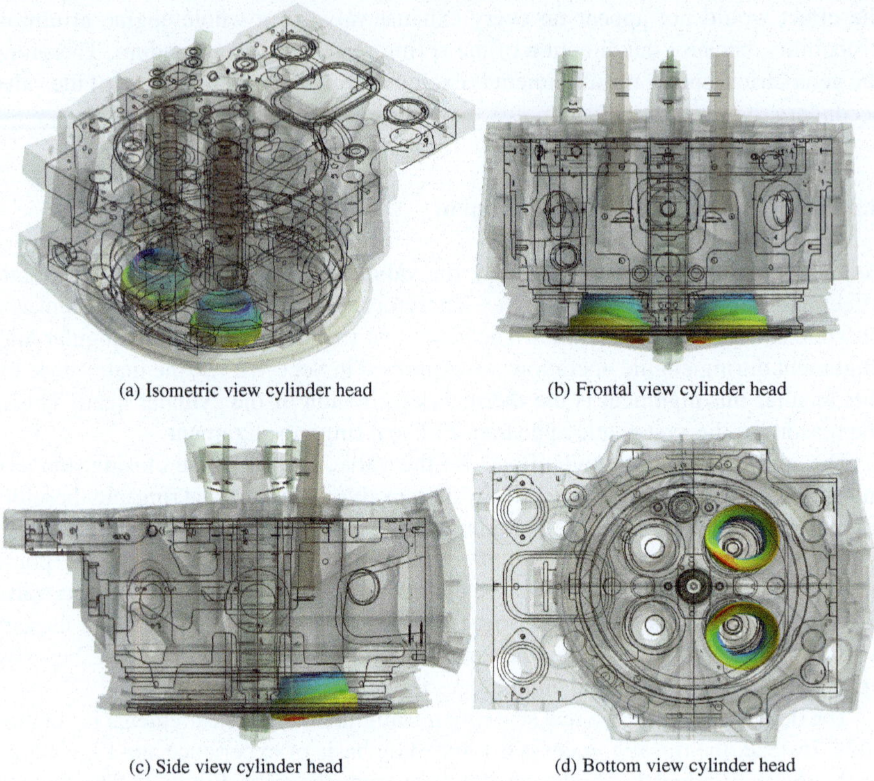

(a) Isometric view cylinder head

(b) Frontal view cylinder head

(c) Side view cylinder head

(d) Bottom view cylinder head

Fig. 4.30 Valve seat ring deformation during engine operation

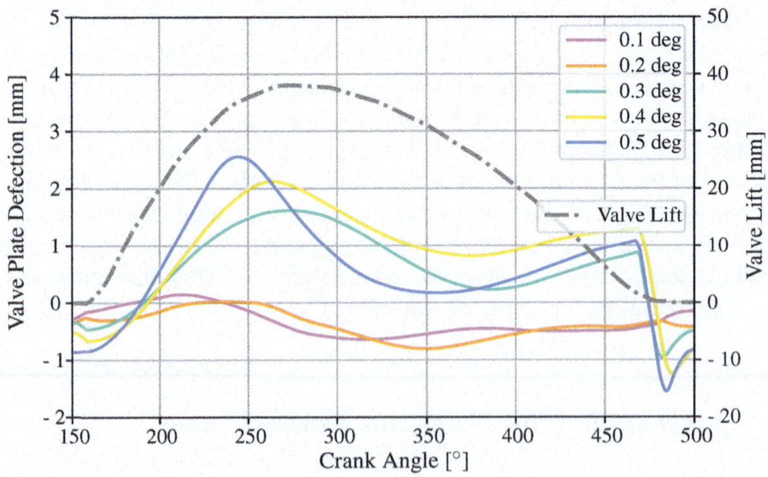

Fig. 4.31 Influence of valve seat ring misalignment

excitation by the transverse component of the flow forces. On the simulative side there is the option of either a static or a dynamic CFD analysis of the flow forces during valve lift to evaluate this hypothesis.

The static analysis provides the amplitude of the occurring transverse forces and requires less simulation effort compared to the dynamic CFD analysis. The CFD simulation results for previously defined boundary conditions are provided by Optimum One GmbH as part of a joint project.

The static pressure on the exhaust side is the boundary condition of the model. The simulation iterates the mass flow on liner side shown in Fig. 4.32 with the aim to reach the given static pressure on the exhaust side. The simulation evaluates the flow forces for five valve positions during valve lift. Table 4.3 shows the resulting forces for all valve positions. The definition of the coordinate system is consistent with the MBS model, therefore the axial valve direction is the y-direction with the positive direction pointing to the valve stem top face. The x-direction points to the liner centre, the z-direction results from the definition of x- and y-direction in the right-handed Cartesian coordinate system. The highest force acts in axial valve direction. This is

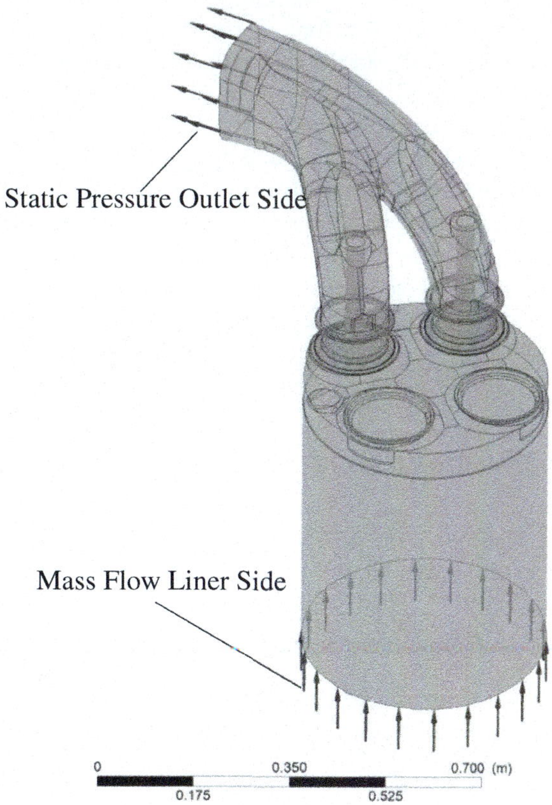

Fig. 4.32 Boundary conditions of the static CFD analysis on 45/60 exhaust valves

Table 4.3 Flow forces on exhaust valves 45/60

Valve lift (%)	Forces valve 1			Forces valve 2		
	x [N]	y [N]	z [N]	x [N]	y [N]	z [N]
2.5	–91	15646	–39	–59	15626	22
6.0	–67	13819	–67	–46	13940	154
12.0	–55	12224	–137	–75	12474	150
50.0	–492	9461	–209	–300	8551	222
100.0	–14	625	–23	–7	612	19

plausible, because this is the main flow direction of the exhaust stream. This force is already implemented by the uniaxial force resulting from cylinder pressure on the valve plate bottom.

The force components in transverse directions are much smaller. The maximum transverse force component with an amplitude of 492 N appears at valve 1 in x-direction at 50% valve lift. The characteristics of the force components between the discrete valve lift positions is not known in the static analysis. By applying the maximum transverse force to the OVM observed at 50% valve lift no excitation of the valve similar to the results from measurements can be obtained. Additionally, the FEM based analysis of the slimmer 48/60 valve already showed that a static force able to displace the valve plate far enough to reach the flame ring is about 4400 N and therefore multiple times higher than the simulated transverse flow force components. A consecutive step regarding investigation of flow forces is the much more resource-consuming transient CFD simulation. To ensure the necessity of this costly analysis, the possible mechanism of valve excitation is investigated in more detail. The open question in this context is, if the valve can be excited dynamically with a much smaller force amplitude than in the static case. If such a excitation mechanism can be found, a transient CFD analysis is a reasonable analysis.

4.3.4 Modal Analysis

The mechanism resulting in vibration is the excitation of eigenmodes of the valve. One of the hypotheses on the excitation of these valve eigenmodes is called arrow theory at MAN ES, because it is derived from a phenomenon observed on an arrow: The axial force transmitted from the bowstring to an arrow during the shot excites the arrow bending mode causing bending vibrations of the arrow during its flight phase. The arrow theory transfers this mechanism to the valve. Within this hypothesis, the axial force of up to 50 kN transmitted by the rocker arm during valve lift excites valve bending modes.

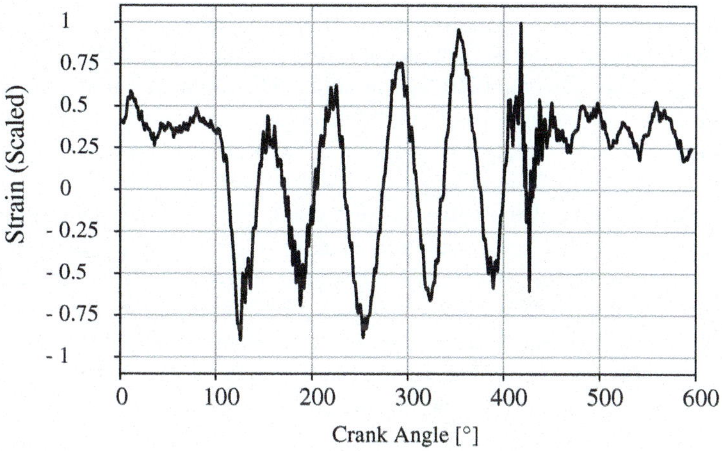

(a) Strain signal valve stem middle

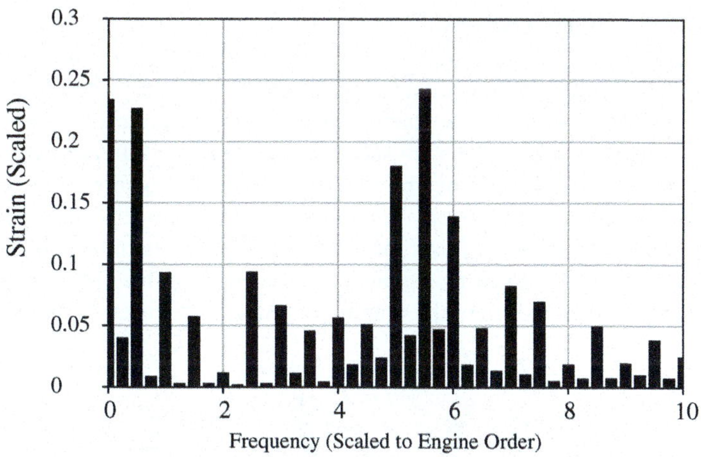

(b) Strain frequency spectrum valve stem middle

Fig. 4.33 48/60 valve bending vibration analysis

If the excitation of the valve eigenmodes is the cause for the bending vibration, the bending frequency observed in measurements and the eigenfrequency of a valve bending mode would match.

For the determination of the real valve bending frequency the strain signal of a 48/60 valve is analysed in the frequency domain. This analysis uses a strain signal from the valve stem middle because the bending strain during the valve lift is assumed to be higher at this position than in the radius. The strain signal of two full working cycles is shown in Fig. 4.33. The adjacent diagram shows the corresponding amplitude spectrum. Each bar in the diagram represents a certain mode of the valve. All modes are transformed to engine orders. The first dominant mode of the amplitude spectrum appears at the 0.5th engine order. This is plausible for a four-stroke engine, because the exhaust valves are actuated in every second crankshaft rotation making the actuation of the valve the 0.5th engine order. The second dominant area is in the area of the 5–6th engine order with the highest amplitude ratio for the 5.5th engine order. This order is not related to any known excitation on the engine. It does not correspond to the cylinder number or a camshaft excitation order.

The modal analysis of the valve delivers results to compare to this strain signal amplitude spectrum. The component modal analysis of the valve is performed in the Ansys Workbench module *Modal*. All eigenfrequencies presented in this section are scaled to specific nominal speed engine orders. The following subsections explain the simulative modal analyses delivering modes to compare to the measurement results. Beside the modal analysis of the valve component a modal analysis of the OVM system is performed. Therefore, the system modal analysis is performed in the MBS model.

4.3.4.1 Valve Modal Analysis

The eigenmodes and eigenfrequencies depend on the boundary conditions. The real boundary conditions of the valve change during one working cycle. Boundary conditions simplified for the modal analysis are used. There are different scenarios with specific boundary conditions. The first analysis obtains the eigenmodes of the unsupported valve. The first six modes are rigid-body modes because the unsupported valve has all six DOF. Figure 4.34 shows the first six linear-elastic mode shapes that are all bending modes. The resulting frequencies, scaled to engine orders and shown in Table 4.4, are much higher than the dominant frequencies observed in the measure-

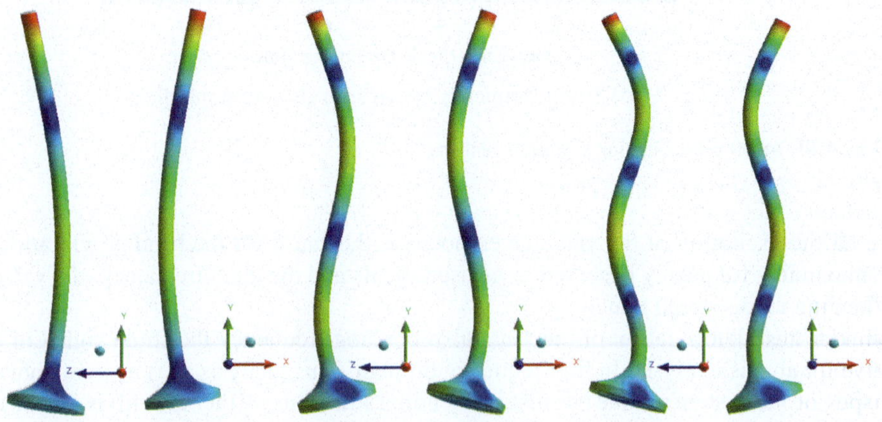

(a) 1. Elastic bending mode (b) 2. Elastic bending mode (c) 3. Elastic bending mode (d) 4. Elastic bending mode (e) 5. Elastic bending mode (f) 6. Elastic bending mode

Fig. 4.34 Elastic mode shapes of unsupported valve

Table 4.4 Elastic modes of unsupported valve

Elastic mode	Frequency (scaled w.r.t. engine orders)
1	18.36
2	18.36
3	53.23
4	53.23
5	103.70
6	103.70

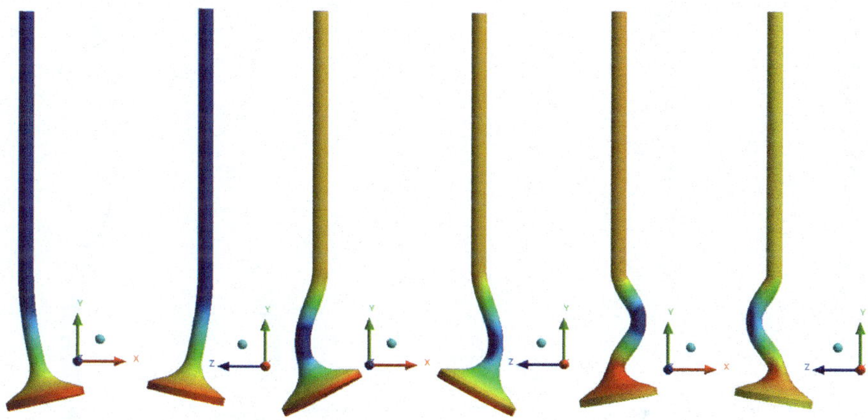

(a) 1. Elastic bending mode (b) 2. Elastic bending mode (c) 3. Elastic bending mode (d) 4. Elastic bending mode (e) 5. Elastic bending mode (f) 6. Elastic bending mode

Fig. 4.35 Elastic bending mode shapes of cylindrical supported valve

ments (see Fig. 4.33). The dominant modes in the measurement are around the 5.5th engine order while the eigenmodes of the unsupported valve start from approx. the 18th engine order.

For the next modal analysis boundary conditions are added to the valve for a more accurate approximation of the valve built-in state. Therefore, a fixed support is added to the valve guidance area. Two points of the overall valve working cycle are analysed. The two states are the almost closed valve and the fully open valve at maximum valve lift. The two states differ in the position of the fixed support. When the valve opens, it moves downwards, the guidance stays in its position in the cylinder head. Hence, the guided area of the valve moves to the upper part of the valve for an open valve. Figure 4.35 shows the first six bending modes. As the mode shapes of the two analysed states look similar, apart from the position of the fixed support, only the mode shapes of the almost closed valve state are shown. Table 4.5 shows the corresponding frequencies for both analysed states [6].

Table 4.5 Elastic modes of cylindrical supported valve

Elastic mode	Frequency (scaled) closed length	Frequency (scaled) full lift length
1	11.48	9.27
2	11.48	9.27
3	110.38	92.33
4	110.40	92.35
5	402.26	306.89
6	402.39	306.98

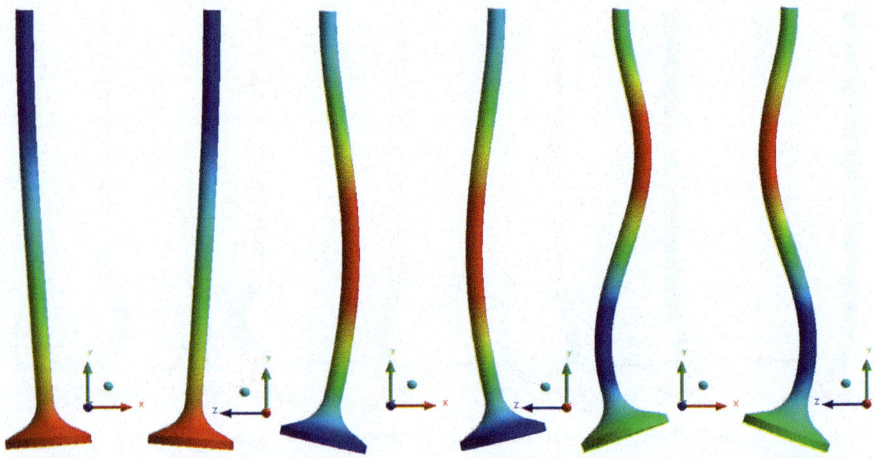

(a) 1. Elastic bending mode (b) 2. Elastic bending mode (c) 3. Elastic bending mode (d) 4. Elastic bending mode (e) 5. Elastic bending mode (f) 6. Elastic bending mode

Fig. 4.36 Elastic bending mode shapes of top fixed valve

The bending modes appear in both transverse directions and therefore all frequencies appear twice. The small deviations in frequencies between the corresponding modes result from a valve mesh that is not perfectly rotational symmetric or other small numerical inaccuracies.

In general, the full lift state model shows the lowest frequencies but even the first bending mode has a frequency of the approximately 9th engine order and is therefore much higher than the observed orders in the measurements. The assumed boundary conditions do not consider the guidance clearance.

Therefore, an analysis similar to the modal analysis in [4] is performed with the fixed position at the valve top. This valve state refers to the situation where the valve is guided by the rocker arm due to the contact between elephant foot and valve top and the valve is ideally centred in the guidance and has no contact point with it (Fig. 4.36).

Table 4.6 Elastic modes of top fixed valve

Elastic mode	Frequency (scaled w.r.t. engine orders)
1	1.92
2	1.92
3	18.72
4	18.73
5	54.70
6	54.71

For this analysis the overall result is similar to the previous analyses, because the frequency range observed in measurement data can not be found in the modal analysis results. In contrast to the previous results the first bending mode frequency is below the measurement data frequency (Table 4.6).

In summary, the valve modal analysis shows no accordance with the bending vibrations obtained in the measurement data. Even for varying boundary conditions, no accordance of the frequencies can be obtained. Therefore, the hypotheses implying the excitation of valve eigenmodes are falsified by this analysis. This affects not only the previously mentioned arrow theory, but all theories based on the assumption of a valve eigenmode excitation.

4.3.4.2 Valve Train System Modal Analysis

In reality, the boundary conditions are more complex. To consider these complex interactions of the components, the excitability of the valve train as a system is investigated. Therefore, this section applies not only to a single valve but the valve train in form of the OVM. In this context, the analysis considers the guidance clearance and the changing contact points in the contact area of valve and guidance, respectively. The valve springs, which also have a high eigendynamic and may influence the overall valve train dynamic, are also considered. The MBS approach allows contacts with clearance due to the state space description of the system. Moreover, other influencing components like the spring with its high dynamic are already included in the OVM. This modal analysis of the valve train system examines whether the excitation of the valve train system in contrast to the excitation of the single valve component has different eigenfrequencies and eigenmodes including a bending of the valve.

Before performing a system modal analysis, an investigation on the response behaviour of the OVM for the example of a 48/60 valve train is carried out to support the hypothesis of system modal excitation. For this purpose a transverse force is applied to the valve plate. The force signal is a sinus signal with an amplitude of 200 N (peak-to-peak of 400 N) and a frequency of the 5.5th engine order. This is the dominant order of valve strain signal obtained from the measurements. The

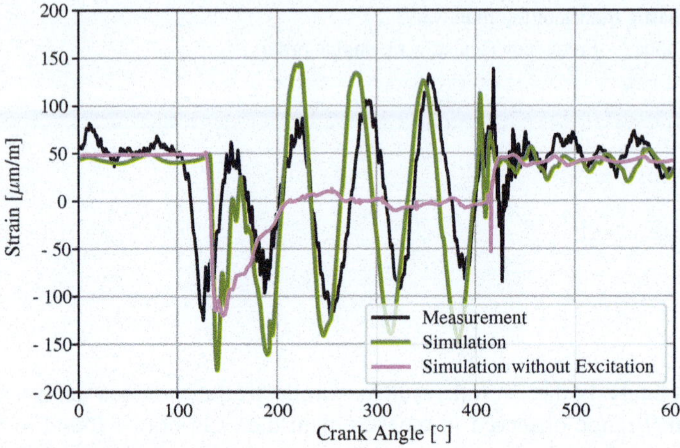

Fig. 4.37 Comparison of measured bending strain, simulated bending strain with generic excitation and simulation without excitation

choice of amplitude is based on the static CFD analysis that showed transverse force components in a range of 7–492 N (Table 4.3). The amplitude for the simulated excitation force is approx. the mean of these force levels. The applied amplitude is much smaller than the static force amplitude necessary to deform the valve enough to reach the observed deformations (see Fig. 4.28).

The strain signals of measurement and simulation, shown in Fig. 4.37, have a good accordance with each other. As a cross-check the simulation is run without the transverse force boundary condition. The results (magenta curve in Fig. 4.37) show that the strain amplitude is much lower without the forces. In conclusion, the exciting force has a significant influence on the valve dynamic although the amplitude is very small.

The next step is finding the reason for the excitability of the valve train by small dynamic forces. For this purpose the eigenmodes of the valve train system of the OVM are determined.

The MBS tool uses the eigenvalue solver for this analysis. The theoretical background of the eigenvalue solver is explained in Sect. 2.1.3. The linearisation of the equations of motion is only accurate for equilibrium states of the model. Therefore, the equilibrium state must be generated before performing the eigenvalue analysis. The modal analysis examines the system's eigenmodes for several valve positions during valve lift.

For the purpose of equilibrium state determination the non-linear method of position adjusting is enabled (see Sect. 2.1.5.1). The steps of the process of the system modal analysis are the following:

1. Start of a regular time integration for the OVM including all loads and the valve lift.

Table 4.7 Bending eigenmodes of the OVM

Valve lift (ratio of max. lift [%])	1.3	13.1	39.4	65.6	100
Eigenmodes	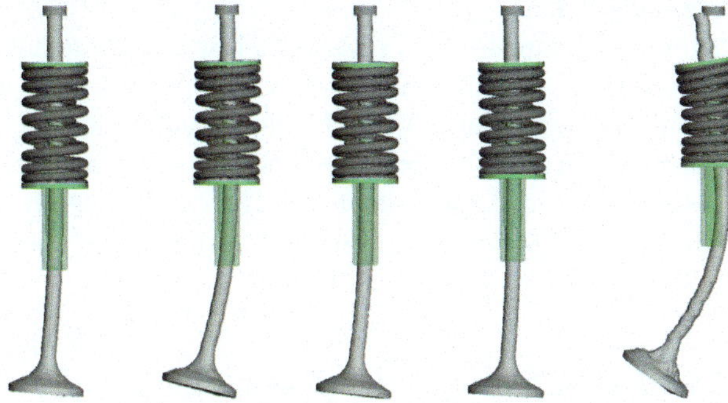				
Frequency (scaled to engine order)	6.1	6.2	6.3	6.2	6.6

2. Stop the time integration at a time t corresponding to the valve position to be investigated.
3. Calculate the equilibrium state at this position to eliminate system velocities and accelerations.
4. Start the eigenvalue solver to obtain eigenfrequencies and the corresponding eigenmodes.

By following these steps the analysis is performed for five valve lift positions.

The results show that the OVM has bending eigenmodes with frequencies in the range of approx. The 6–6.5th engine order for each of the investigated positions. All bending mode shapes shown in Table 4.7 are scaled by the same factor. Modes with higher kinetic energies, e.g. the eigenmode at 100% lift, show higher deformations.

The frequencies are a little bit higher than the measured frequencies but very close to the observed frequency range.

A bending of the valve can be observed for all valve lift positions. This finding explains the results shown in Fig. 4.37: The valve train has eigenfrequencies in the frequency range observed in the strain measurements. The excitation with a frequency similar to eigenfrequencies leads to a vibration of the valve train. The valve train vibrations show a bending of the valve. The observed eigenmodes are modes of the valve within the valve train as a system including the contacts and spring dynamic. In summary, the results of the presented investigations lead to three main findings:

1. The bending vibration of the valve is not an eigenmode of the valve as a separate component, but an eigenmode of the valve train. The eigenfrequencies of the valve train for various valve lift positions are close to the frequency range observed in measurements.
2. Although a static force of several kN is necessary to deflect the valve plate enough to reproduce the valve plate—flame ring contact, the necessary force amplitude decreases significantly for a dynamic excitation. The observed valve secondary dynamic can therefore be traced back to the excitation of valve train eigenmodes. A relatively small force amplitude can excite the valve eigenmode.
3. The origin of the exciting force is not known and needs to be figured out.

4.3.5 Excitation of Valve Train Eigenmodes

The response analysis in Sect. 4.3.4.2 shows that the excitation force must reach amplitudes in an approximate range of 200 N to result in valve vibrations with the observed magnitude, and is likely to be continuous, because the measurement shows bending strain during the whole valve lift.

The OVM already includes the valve actuation with the force transmitted from rocker arm to valve, the valve spring dynamic and the form and amplitude of the cylinder pressure. This cylinder pressure is taken directly from the measurements, the spring force and the pushrod force that directly lead to the rocker arm force can be verified by measurement data. Because all these considered forces do not excite the OVM, it follows that there must be a further force acting on the valve train system.

The analysis in Sect. 4.3.3 indicates amplitudes of transverse flow forces that are high enough to excite the valve train. However, the analysis is static while dynamic excitations are required here. The response analysis and the modal analysis of the OVM give a clear indication of a dynamic excitation. The magnitude and characteristics of the continuous flow forces have to be obtained to investigate if these excite the valve train. The following investigations are carried out for the 49/60 valve train.

4.3.5.1 Implementation of Dynamic Flow Forces in MBS Model

The generation of a continuous flow force signal requires a dynamic CFD analysis of the exhaust cycle. The department of thermo- and fluid dynamics at MAN ES performs this CFD analysis for the nominal load point and provides the results in form of pressure information for coordinates of the discretised combustion chamber including the exhaust valves. The applied turbulence model is the common $k - \epsilon$ model [2, 7].

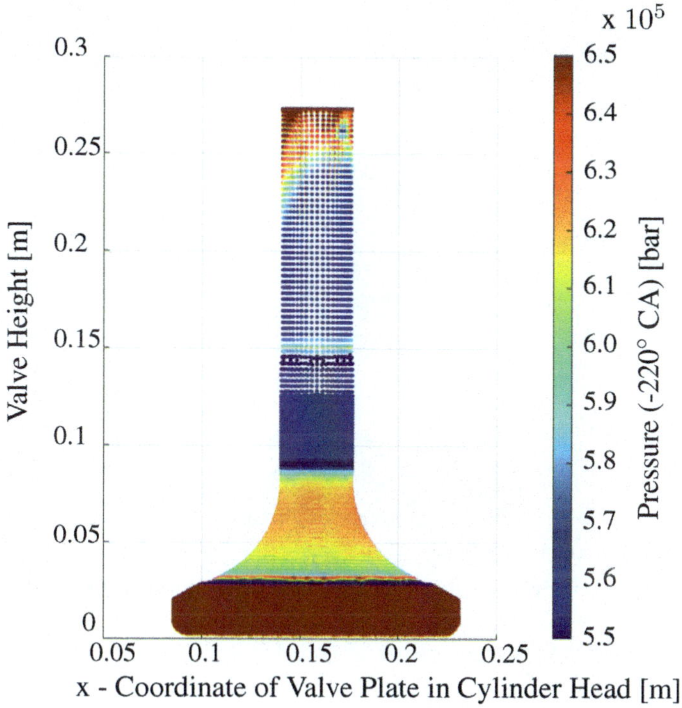

Fig. 4.38 Raw pressure data on valve

Focusing on the valves, the pressure of the exhaust stream acts only on the area of the valve stem and plate below the lower edge of the valve guidance. Figure 4.38 shows the raw coordinate pressure data for one of the exhaust valves.

The MBS model needs forces as input data. Moreover, these forces can only act on master nodes which have to be defined in the reduction process of the flexible body. In conclusion, changes in the OVM and also in the CFD data are required. The pressure data needs to be transformed to force input data. Master nodes have to be added to the valve flexible body for the application of these force input data.

For the conversion of the pressure data to force data multiple options are possible. The simplest way is the transformation of the pressure data of all coordinates that belong to the valve to one resulting force vector. Because the valve is a flexible body, this simplification could distort the effect of the flow force on the valve. Thus, the pressure load is transformed to a quasi-line load: The valve stem is discretised into small horizontal sections. Each section has a master node in its centre. Hence, a three-dimensional force vector can be applied to each section. The force vectors of each section are independent. The discretisation of the valve stem defines the number of additional required master nodes on the valve.

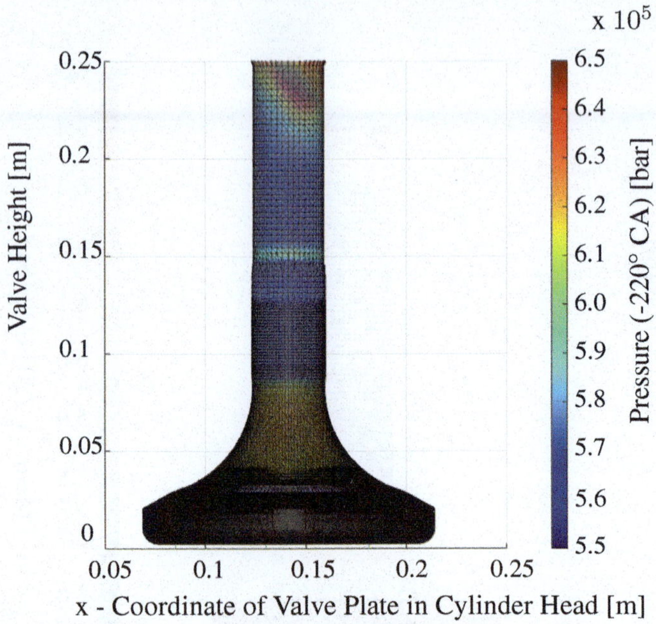

Fig. 4.39 Triangulation valve hull

The transformation of the pressure data to force vectors requires the determination of a corresponding surface and the surface normal direction. As mentioned above, the CFD simulation delivers data for coordinate points on the valve surface. By generating a mesh, the surface of the valve can be determined. The algorithm used to generate a mesh from the coordinate points is the Delauney-triangulation. This algorithm generates surface elements from the coordinate points. The surface elements always have a triangular shape. Therefore, three coordinate points build a element of the mesh. The resulting mesh is shown in Fig. 4.39. The Delaunay-triangulation ensures that none of the given points is inside the circumference of any of the triangles. That way the minimum angle of every triangle is maximized. This leads to a minimization of rounding errors of the discretisation [1].

For each of the triangles the surface normal is obtained in order to determine the resulting surface normal of each section on the valve. The surface normal of each valve stem section is the resulting normal from each triangles surface normal belonging to that section.

The pressure data is given in steps of two degree crank angle and results in force vector components depending on the crank angle or the time, respectively, and the valve height. The force components in the valve plate sections are higher than the force components in valve stem section. The reason is that the valve surface, which the pressure acts on, is bigger at the valve plate sections than the area at the valve stem sections. Figure 4.40 shows the force characteristics for all directions. The crank

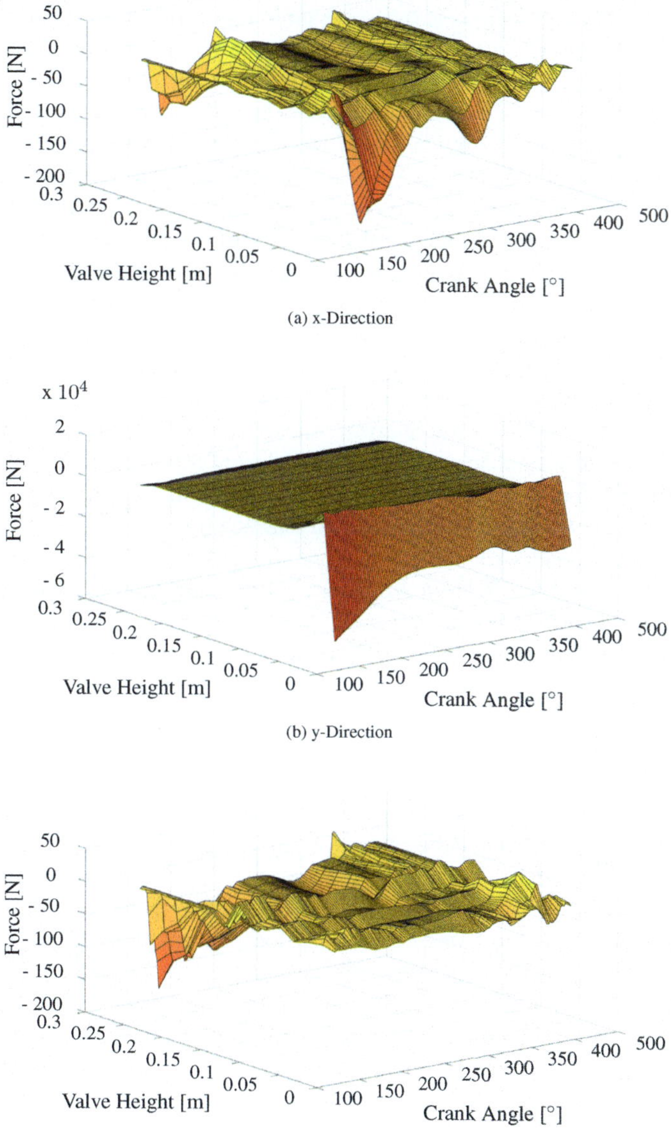

(a) x-Direction

(b) y-Direction

(c) z-Direction

Fig. 4.40 3D-flow force components exhaust valve

angle axis in the plots shows the force characteristics in dependency of time, the valve height describes the force characteristics in dependency of the axial valve coordinate: The point of origin on this axis is the valve plate bottom. The upper point is on the valve stem right below the valve guidance. This is the valve area affected by the flow forces. The force axis gives the force amplitude for all directions, the sign is consistent with the previously defined valve coordinate system. The force component in the axial or y-direction is much higher than in the transverse directions (x- and z-direction). The axial force on the valve acts on the valve plate bottom. This result is expected, because the main flow direction from the cylinder towards the exhaust channel and the exhaust pipe is mainly in axial valve direction. This force corresponds to the cylinder pressure that was already applied to the valve plate (see Sect. 4.2.1.1).

Figure 4.40 shows the force components not vibrating symmetrically but having a shift towards the negative x- and z-direction. Furthermore, the force development over time indicates a fluctuation of the transverse load over the whole valve lift period with decreasing amplitudes.

In comparison to the response analysis, shown in Fig. 4.37, the amplitude of the transverse forces is not constant but decreases during the valve lift. With this input data, the simulation results can directly be compared to measurement results, which is a major innovation regarding valve train dynamic simulation.

The simulative results for the subsequent simulation-measurement comparison and simulation model verification are generated by including time-depending flow force vectors to the OVM model and implementing the valve flexible body with the adjusted reduction and subsequently solving the model with the new setup.

4.3.5.2 Validation of the OVM

The basis for the simulation model verification is the comparison of the strain signals resulting from the simulation of the valve working cycle and the measured valve strain signal.

The measurements provide signals in the valve plate radius and the valve stem middle. The comparison uses the data of the nominal operation point of the 49/60 engine that is 100% load and 600 rpm. The simulation valve clearance is taken from the measurements (see Sect. 4.1.5). The input cylinder pressure is the transient mean cylinder pressure from the measurements, the transverse flow forces are taken from the CFD analysis. A seat angle misalignment of 0.1° is assumed. The direction of the misalignment corresponds to the thermally induced cylinder head deformation. The measurement results are examined from measurements described in Sect. 4.1. Therefore, it is possible to compare results for both strain measurement positions at the valve. For comparison, the three-sigma intervals of the measurement values are taken to include statistical variations in measurements during the whole valve lift phase.

Figure 4.41a shows the comparison of simulation and measurement data for the valve lift phase at valve stem middle sensor position. The magnitude of the simulation curve and the overall characteristics of the strain in the stem middle is very similar to

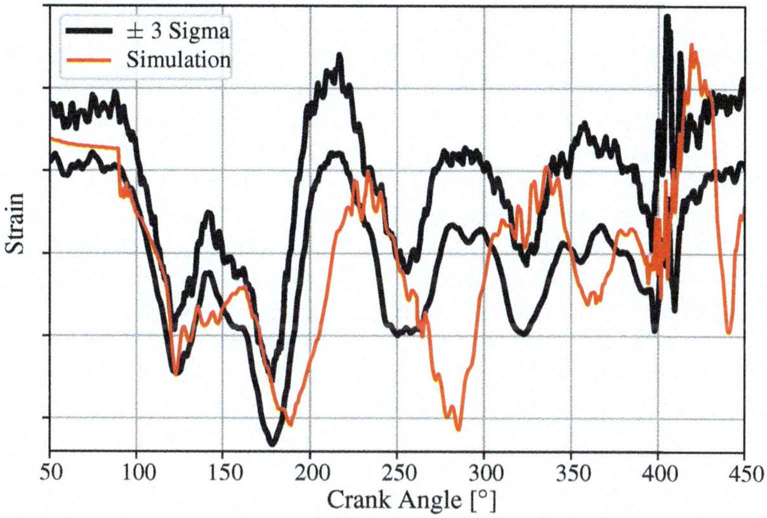

(a) Valve stem middle, exhaust valve 1

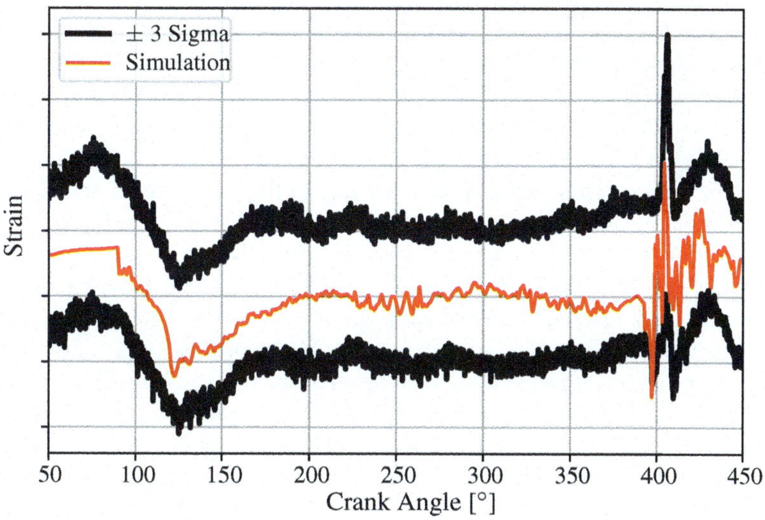

(b) Valve plate radius, exhaust valve 1

Fig. 4.41 Comparison of valve strain data simulation versus measurement the measurement. The main difference between the curves is a phase shift between the simulation and the measurement. This phase shift indicates a lower system stiffness of the simulation model in comparison to the real valve train. There are a few possible reasons for the lower stiffness of the simulation model: The contact stiffness of the valve-guidance contact can not be properly determined and might even change during engine operation. The guidance has no own lubrication supply which makes it impossible to determine the amount of lubricant in the guidance. A further possible reason is the simulation of the coil-to-coil contact of the valve spring. This contact is discretised in the simulation model, in reality the coils have a continuous contact. Thus, the contact stiffness and damping differs between simulation and reality and influences the overall system stiffness. The spring analysis in Sect. 4.2.1.4 already indicated a slightly lower frequency of the springs compared to the ideal spring stiffness. Another potential influence of the valve vibration frequency is the frequency of the flow force vibration. As the flow force data also result from simulation a deviation of the real flow force vibration frequency and the simulated flow force frequency is possible. This slight phase shift is acceptable for the accuracy of the simulation model as it does not distort the overall dynamic and load on the valve.

Secondly, the strain data of the valve plate radius is compared. Figure 4.41b shows the results of this comparison. The first peak in the strain characteristics at approx. 120° crank angle is the normal strain in the whole valve caused by the actuation of the valve by the elephant foot. In the further progression of the strain a bending can not be observed in the radius. This is plausible for the 49/60 exhaust valve because it has a large diameter in the area of the plate-to-stem radius and therefore a high bending stiffness. The valve bending is located in the stem.

Once the valve closes, a strain peak appears in the strain signal at approx. 400° crank angle. The simulation data follows the measured strain data progression very accurately. The strain amplitude during valve closing lies in between the upper and lower borders of the measurement data and is therefore a good approximation of the overall mean peak load during valve closing.

A cross-check is performed to verify the influence of flow forces. For this cross-check the flow forces are switched off in simulation. The results in Fig. 4.42 emphasize the drawn conclusions: Without the flow forces the strain in the valve stem differs significantly. The amplitudes during the valve opening phase are much smaller. Moreover, the strain characteristics shows almost no accordance of simulation and measurement.

The comparison of the cross-check simulation and measurement data in the valve plate radius shows similar characteristics during the valve opening phase. This result is expected, because the bending in the valve plate radius is small because of the higher bending stiffness in this area compared to the valve stem. Nevertheless, the valve closing indicates a significant difference between the simulation results generated with flow forces and the simulation data generated without flow forces: The bending peak-to-peak load during valve closing is almost twice as high when considering the flow forces and therefore fits the measurement results significantly better than the resulting strain peak from the simulation without flow forces. This is an unexpected result because the amplitude of the flow forces is small and was therefore not expected

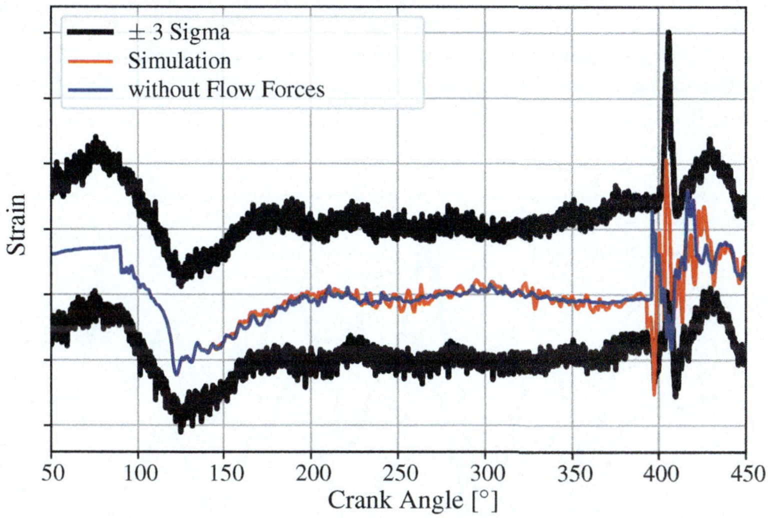

(a) Valve stem strain, exhaust valve 1

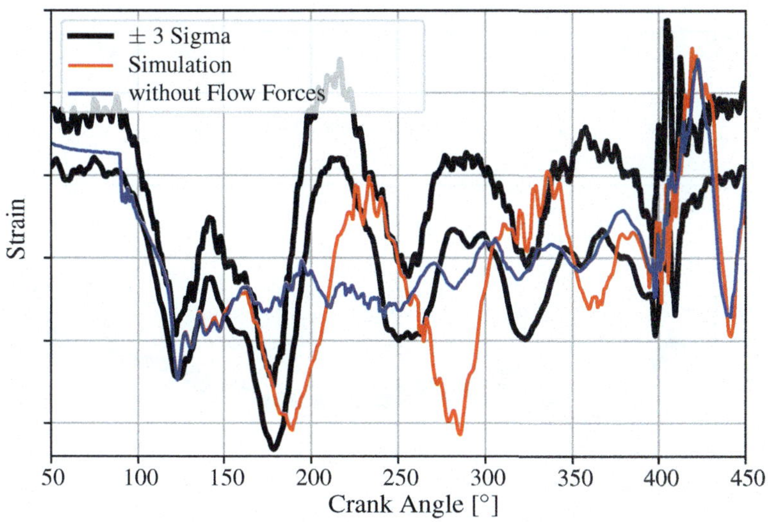

(b) Valve plate radius strain, exhaust valve 1

Fig. 4.42 Comparison of strain result without flow forces

to have such a great impact on the valve. The inclusion of flow forces is a new approach regarding the simulation of valve dynamic.

In conclusion, the major finding of this analysis is that including flow forces to the OVM leads to model able to describe the valve dynamic sufficiently. Therefore, the valve loads resulting from this dynamic can also be simulated sufficiently, enabling a predictive determination of load side of the fatigue strength assessment. In order to obtain accurate valve load data for the heavy-duty exhaust valve, the implementation of dynamic flow forces is mandatory. The validation also includes the second exhaust

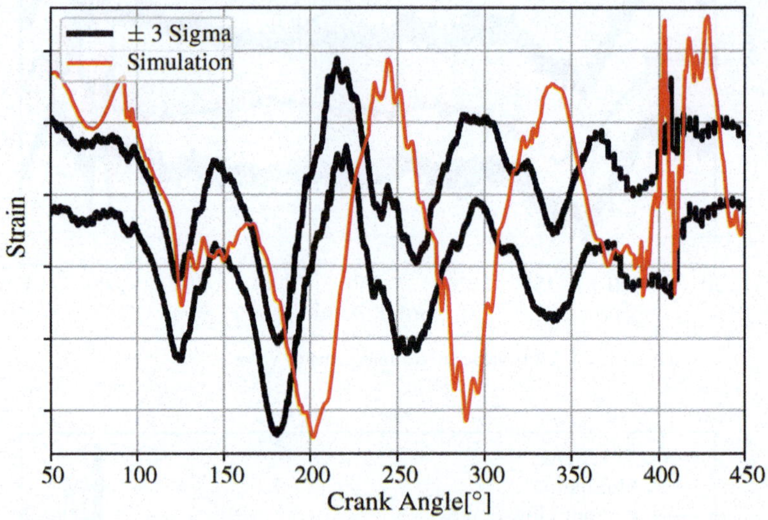

(a) Valve stem middle strain, exhaust valve 2

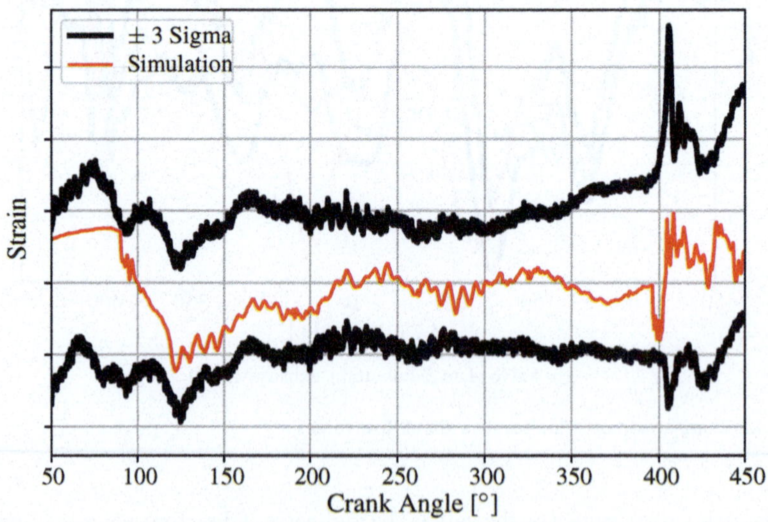

(b) Valve plate radius strain, exhaust valve 2

Fig. 4.43 Comparison of strain results, valve 2

valve. The validation procedure is the same for both exhaust valves. Both evaluation positions show similar accordance with the measurement data. On simulation basis, the only difference between both models is the amplitude and direction of the transverse flow forces. Figure 4.43 gives an overview on the validation results of valve 2. The load of valve 2 is in general lower, the strain peaks are smaller compared to valve 1. Therefore, the fatigue strength assessment focuses on valve 1 as the higher loaded component.

The findings and analyses in this chapter lead to a validated simulation model including all relevant physical effects to describe the valve dynamic. The subsequent chapter builds a fatigue strength assessment based on this simulation model.

References

1. FISHER, M. ; SPRINGBORN, B. ; SCHRÖDER, P. ; BOBENKO, A. I.: An Algorithm for the Construction of Intrinsic Delaunay Triangulations with Applications to Digital Geometry Processing. In: *Computing* 81 (2007), Nr. 2, 199–213
2. LAUNDER, B. ; SPALDING, D. : The numerical computation of turbulent flows. In: *Computer Methods in Applied Mechanics and Engineering* 3 (1974), Nr. 2, 269–289.
3. LEWIS, R. ; DWYER- JOYCE, R. : *Automotive Engine Valve Recession*. Professional Engineering Publishing, 2001 (Engineering Research Series (REP))
4. MESSER, M. : *Tribologische Untersuchungen des Systems Ventilsitz - Ventilsitzring*. Hannover, Gottfried Wilhelm Leibniz Universität, Dissertation, 2013
5. PAUSIN, A. ; BECK, A. ; BÖHM, P. ; SPECKERT, M. ; FIEDLER, J. : Ventiltriebsmessungen von Großmotoren zur Lastermittlung. In: *DVM-Workshop Zuverlässigkeit und Probabilistik* (2021)
6. PAUSIN, A. ; BECK, A. ; BÖHM, P. : Simulativer Ansatz zur Abbildung der Ventilsekundärdynamik von Großmotoren. In: *NAFEMS DACH virtuelle Konferenz 2020* (2020)
7. VERSTEEG, H. K. ; MALALASEKERA, W. : *An introduction to computational fluid dynamics - the finite volume method.* Pearson education, 2007

Chapter 5
Application

The previous chapter delivered a simulation model of the valve dynamic, able to reproduce measurement data and give a representation of the real valve dynamic and valve loads.

The present chapter outlines the application of this model. The main aim of the application is to determine the load data for the valve fatigue strength assessment. The fatigue strength assessment relates the actual load of the critical spot of a component to the endurable load. The chapter shows the determination of both parts of this fatigue assessment. The first part of the chapter focuses on the side of the actual loads. Thus, studies of parameter sensitivity towards the valve strain and valve dynamic deliver information over the expectable load collective. In this context, the DoE provides methods to determine the magnitude of influence of several factors on the model outputs. With respect to the aim of fatigue assessment the most relevant output is the component load.

With respect to the endurable component load, the chapter also describes the material data determination, the determination of all relevant input data for the fatigue assessment and the determination of the relevant amplitude of valve load. Finally, the chapter shows the results of the final fatigue assessment.

5.1 Simulative Load Estimation

The evaluation of valve train measurements delivers the distribution of maximum valve strain for one cylinder of one engine type. For the reason of extremely high costs, measurement data for more than one cylinder or for several engines of the same type rarely exist.

A. Eret, *Dynamic and Fatigue Assessment of Heavy-Duty Engine Valves*, Mechanics and Adaptronics, https://doi.org/10.1007/978-3-031-49151-1_5

Nevertheless, an estimation of the expected load spectrum is necessary for the fatigue strength assessment. The valve load depends on several influence factors. Many influence factors are known. However, the magnitude of the influence is not clear. Thus, the factors, which are parameters of the valve train model, and may influence the valve dynamic and maximum valve strain are determined. Twelve influencing parameters can be identified that are shown below:

- **Guidance Clearance**
  Description: A larger guidance clearance may decrease the guidance wear, but is likely to increase the valve bending amplitude because it leaves larger space for uncontrolled valve movement. The opposite effect appears for a smaller valve clearance. The guidance clearance may vary due to manufacturing tolerances. Another point is the increasing guidance clearance due to wear during its operation time. Furthermore, the temperature-dependent guidance deformation can lead to an increasing guidance clearance, especially in the lower guidance part.
  Implementation: The guidance clearance is a parameter of the pre-implemented EHD functionality of SIMPACK. Therefore, this parameter needs no additional implementation.
- **Seat Ring Angle**
  Description: An angular misalignment of the seat ring can appear due to manufacturing and mounting tolerances. Another relevant cause for an angular misalignment is the temperature related deformation of the cylinder head, especially on the exhaust side (see Fig. 4.30).
  Implementation: Due to the implementation of the valve to valve seat ring contact as a PCM the seat ring geometry is fully implemented. Therefore, the angular misalignment can directly be applied as a rotation of the valve seat ring around one of the transverse valve seat ring axes.
- **Guidance Deformation (Modal Description)**
  Description: The guidance deformations shown in Fig. 4.20 are caused by the temperature related cylinder head deformation. The deformation increases the eccentricity of valve and valve seat ring.
  Implementation: The modal guidance shapes overlay the nominal guidance geometry. A 2D-function describes the first four modal shapes. These first four modal shapes are assumed to influence the contact behaviour. The deviation from the ideal geometry caused by each mode are superposed to the cylindrical geometry. The 2D-functions have the guidance height and the circumferential coordinate as input variables. These functions can be scaled by arbitrary factors for the influence determination. The following Fig. 5.1 shows the deformed guidance profile. The profile is unwrapped along its circumference.
- **Valve Clearance**
  Description: There is a clearance between the valve top face and the elephant foot. During the engine operation, there is a thermal extension of valve train components on the exhaust side. The clearance between valve and elephant foot ensures that there is no preload on the valve train due to the thermal expansion of components.

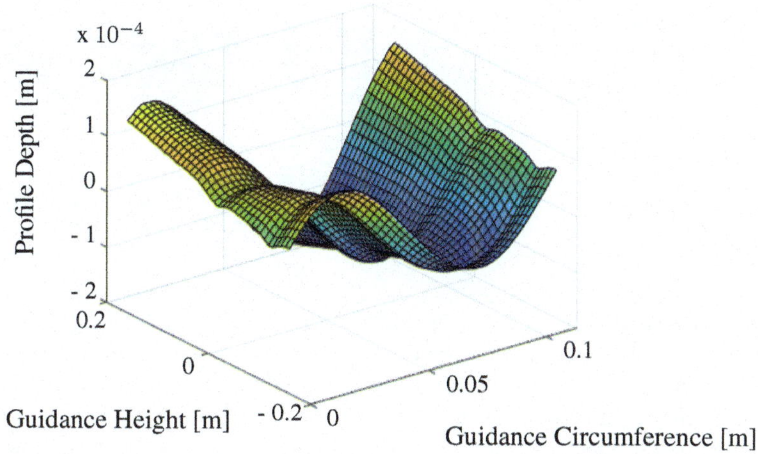

Fig. 5.1 Deformed guidance profile

The size of the valve clearance is also influenced by valve wear and valve seat wear and the accuracy of the adjustment.
Implementation: Between the valve and the dummy body (implemented to model the elastic rocker arm deformation, see Fig. 4.25), there is an initial gap that corresponds to the valve clearance between valve stem top and elephant foot. This gap is variable.

- **Valve Spring Stiffness**
  Description: Because of the wide tolerances of springs in general, the spring forces for the pretension length l_1 and the length l_2 can vary up to 10% and in turn affect the forces on the valve.
  Furthermore, the spring stiffness depending on the wire diameter varies.
  Implementation: The force tolerance is mainly the result of the geometry tolerance. The spring stiffness is higher for bigger spring diameters and a bigger wire diameter. For a fully variable implementation of the valve stiffness the full geometry of the valve spring has to be changed, the complete reduction process has to be performed subsequently. Because of the immense effort of this procedure, the parameter analysis uses a simplification. The geometry of the spring remains the same, but the stiffness matrix is scaled by a factor. This simplification does not represent the spring behaviour accurately but the influence of a change in the stiffness can be investigated properly. Therefore, it can be analysed whether or not a change in valve spring stiffness may influence the valve dynamic.
- **Valve Spring Misalignment**
  Description: As shown in Sect. 4.3.2.1, the spring tolerances can cause an axial valve spring misalignment and therefore introduce transverse forces to the valve shaft. This transverse forces can occur in every direction and affect the secondary dynamic of the valve.

Fig. 5.2 Valve spring plate shoulder

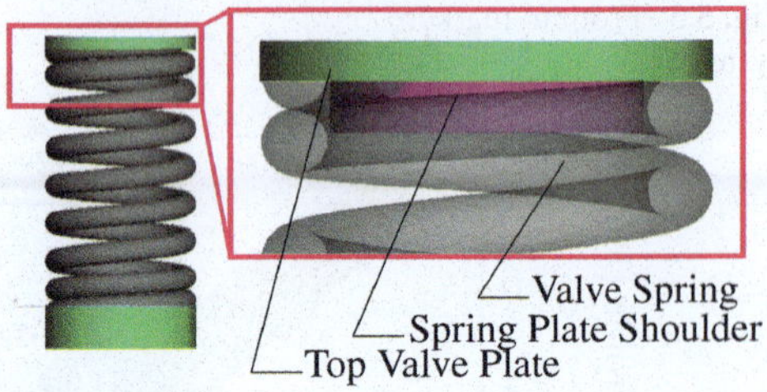

Implementation: The valve spring simulation model consists of the spring, the bottom plate and the top plate. The top plate has a shoulder that centres the valve spring on the top valve plate. If there is a clearance between the inner coil diameter and the shoulder, the valve spring slips on the plate until it reaches an edge of the shoulder and thus gets a misalignment during preloading. The clearance parameter is available in the model and can be modified. Figure 5.2 shows the top valve plate with the shoulder. The green part is the upper valve plate, the grey transparent part are the spring coils. The pink part is the shoulder. It is fixed on the valve plate and centred in the spring coils. The clearance is between the spring plate shoulder and the inner diameter of the spring coils.

- **Transverse Force Amplitude from Elephant Foot**

  Description: The elephant foot should only transmit axial forces to the valve. However, the rocker arm's movement follows a circular path. The elephant foot has a spherical joint to compensate the transverse portion of the rocker arm movement, but due to friction between elephant foot and rocker arm and elephant foot and valve there is a transverse force transmitted to the valve. Moreover, slight misalignment of rocker arm or valve bridge relative to the valve results in a transmission of further transverse forces by the elephant foot.

  Implementation: The force element transmitting the actuation force from the dummy body to the valve gets a transverse component in addition to the axial main component. This transverse component is defined as the axial force component multiplied by a scaling factor.

- **Cylinder Pressure**

  Description: The cylinder pressure depends on deterministic and stochastic influences. It varies on the one hand in dependency of the engine load. On the other hand the cylinder pressure scatters for constant operation points. The varying cylinder pressure influences the force on the valve during ignition as well as the axial and transverse forces on the valve during the exhaust valve opening.

  Implementation: There are measured cylinder pressure characteristics for the engine load points 0%, 50%, 75%, 90%, 100% and 110%. Interpolation of these data delivers the pressure characteristics for all engine load points in between the measured load points.

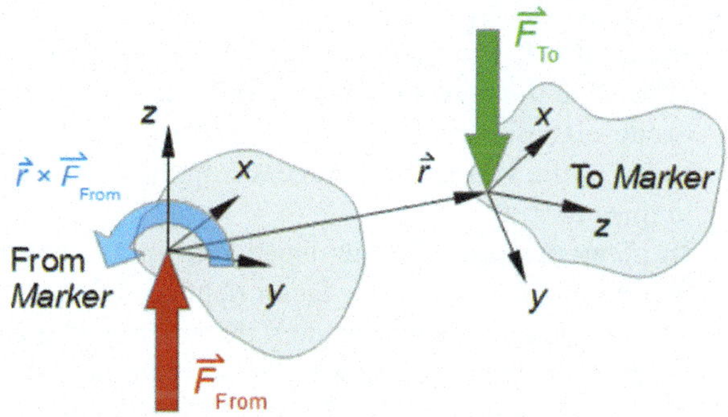

Fig. 5.3 From/to marker systematic, [1]

- **Transverse Flow Force Amplitude**
  Description: The transverse dynamic flow forces can only be determined by CFD simulation. However, a variation of the amplitude of these forces due to scattering or varying engine operation points is expected. Thus, the influence of a varying amplitude is determined.
  Implementation: The functions describing the flow force components are scalable by an arbitrary factor. These functions deliver the input for the time-dependent forces. The range of variation of the flow force amplitude is not known. However, the amplitude depends on the variation of the cylinder pressure at the valve opening. A small margin is added to the cylinder pressure variance to ensure that the variation of force amplitude is not underestimated.
- **Transverse Flow Force Direction**
  Description: Varying engine operation points or a slight shift in the valve opening phase can cause a change in the flow force direction that in turn might affect the valve secondary dynamic.
  Implementation: Force elements apply the flow forces to the valve. They use the from/to-marker systematic shown in Fig. 5.3. The forces are applied in the from-marker system. The from-marker of the force element is defined on the initial system. The rotation of the from-marker therefore changes the direction of the applied flow forces.
  The variation of the flow force direction is not known. There are no simulation results or measurement results available that provide information on this variation. Therefore, the direction variation is assumed.
- **Torsional Vibration Camshaft**
  Description: The torsional camshaft vibration has multiple potential effects. This investigation focuses only on the most dominant effect, which is the shift of valve opening phase.
  Implementation: The implementation of the camshaft torsional vibration uses this simplification of the potential effects. The implementation includes the shift of the cylinder pressure in terms of time relative to the valve lift.

- **Rotational Speed**
  Description: The engine rotational speed varies for different operation points and affects the dynamic of several engine components as some specific speeds might excite eigenfrequencies. The engine rotational speed also affects the actuation speed of the valves. The influence on the valve and the valve train is determined in this context.
  Implementation: All time-dependent input functions are scalable along their time-dependent axes. The scale factor depends on the engine rotational speed.
  This implementation considers the rotational speed influence to be completely isolated. In reality, a slower engine rotational speed comes along with changes in other parameters, e.g. the flow force amplitude or direction and not only a scaling along the time axis. However, the sensitivity study considers effects of these interactions because several parameters are varied simultaneously. Therefore, the isolated consideration of the rotational speed change avoids a doubling of certain effects.

The aim of the subsequent investigation is the determination of the influence of each parameter. All the parameters can be changed script-based and simultaneously. Therefore, sets of parameter combinations can be handed to the OVM via an input data including these parameter combination sets. This enables the full automation of the workflow of parameter evaluation. The tool utilized to implement the workflow (SimWorkflow) is introduced in [12].

The main output value is the maximum strain that is defined as the maximum peak strain during valve closing. The maximum strain is evaluated for the circumference of the valve plate radius, as Sects. 4.1.5.2 and 3.3.3 describe.

The output value taken to describe valve secondary dynamic is the transverse deflection of the valve plate. This output gives no direct information on the valve load and has therefore minor importance. However, the valve plate deflection gives information on the valve dynamic. For the analysis this is defined as the maximum transverse valve plate deflection during the valve opening phase.

An OAT analysis delivers an overview over the general influence of each parameter on these output values. With this information a foundation is built for the choice of an experimental design for the subsequent sensitivity analysis. Therefore, this analysis is performed prior to the sensitivity study. The OAT follows these steps: All parameters except one are kept constant while one factor is varied on multiple levels. This procedure is repeated for each of the factors. A fitting of a linear and a quadratic polynomial interpolation curve to the data evaluates the results. The coefficient of determination R^2_{det} indicates the quality of the fit because it relates the regression sum of squares, also known as sum of squares explained, to the total sum of squares. It is defined by (5.1) [15]. The interpolation polynomial p_{poly} delivers the output values $\hat{Y}_i$. $\bar{Y}_i$ is the mean of the actual outputs Y_i.

$$R^2_{det} = \frac{\sum(\hat{Y}_i - \bar{Y}_i)^2}{\sum(Y_i - \bar{Y}_i)^2} \tag{5.1}$$

$$\hat{Y}_i = p_{poly}(x_{f,i}) \tag{5.2}$$

$$\bar{Y}_i = \frac{1}{n_f}\sum_{i=1}^{n_f} Y_i \tag{5.3}$$

The determination coefficient lies between 0 and 1, while 1 describes the best fit result. Figure 5.4 shows the results of the analysis for the factors flow force angle, translateral elephant foot force on valve, guidance clearance and valve clearance. The x-axis shows the variation range of the parameter, the y-axis shows the result of the output value. The primary output is the valve strain. However, the figure shows the results of the valve plate deflection as well. The results of the simulation are shown by the data points. The regression curves are added to each analysis.

The valve plate deflection often shows a quadratic or, in the case of the guidance clearance dependency (see 5.1.1e), even a linear behaviour towards the influencing parameters. The coefficient of determination is at least 0.8818 for the variation of the transverse force transmitted by the elephant foot (see 5.1.1c).

The output value strain shows a different behaviour. For the variation of the angle of flow forces, shown in 5.1.1b, the strain shows a nearly quadratic behaviour, although the coefficient of determination at a value of 0.7737 indicates a considerably poor fit of the quadratic model compared to the fit of the valve plate deflection output for the variation of the same parameter. The quadratic fit of the strain regarding the translateral elephant foot force or the guidance clearance is poor according to the coefficient of determination (see 5.1.1d).

Table 5.1 shows all results of the analysis for the output value strain and Table 5.2 shows all results of the analysis for the output value valve plate deflection.

The overall results indicate a higher order influence of the factors towards the output, especially regarding the valve strain. This leads to difficulties in the sensitivity analysis, which is explained in Sect. 5.1.2. This analysis gives a first hint on the DoE methods, which are suitable for this parameter evaluation.

5.1.1 *Parameter Reduction*

There are twelve parameters, also referred to as factors in the context of sensitivity studies, potentially influencing the valve train behaviour. This is a relatively large number of influences when interaction effects and higher order influences have to be considered. A full CCD requires 4356 runs for twelve parameters. With a realistic assumption of a 2.5 h simulation time per sample and a parallelisation of two simulations, this approach needs more than 32 weeks to complete the analysis. Because this is unrealistic in terms of practical use, the first step of the sensitivity study of the parameters is the elimination of less important factors with as few simulation runs

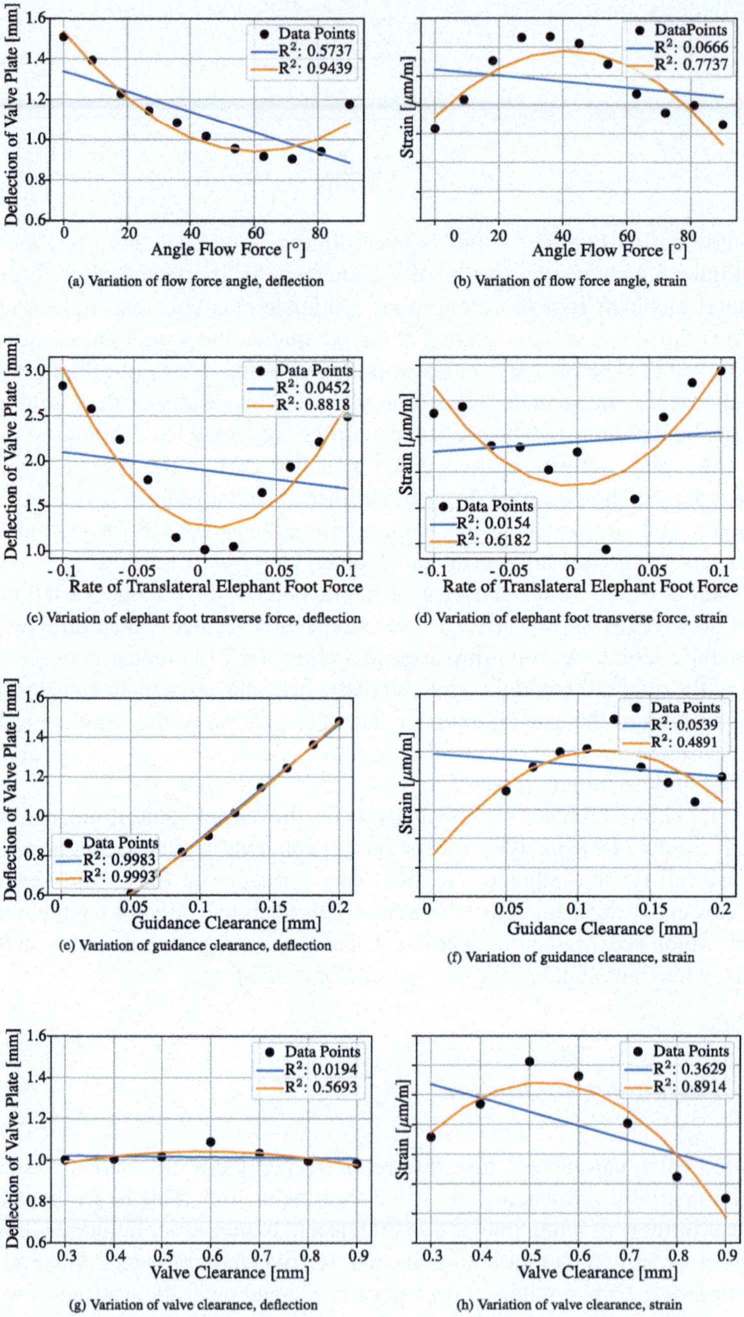

Fig. 5.4 Results of OAT parameter variation

Table 5.1 Evaluation of strain

Factor	Polynom fit		Data range	
	linear	quadratic	Min.	Max.
Guidance clearance	0.0539	0.4891	157.0	228.2
Seat ring angle	0.4981	0.7635	100.4	236.2
Modal guidance deformation	0.6767	0.9594	89.0	228.2
Valve clearance	0.3629	0.8914	112.8	228.2
Valve spring stiffness	0.2921	0.7273	203.0	223.6
Valve spring misalignment	0.2131	0.3628	134.5	236.6
Transverse force amplitude from elephant foot	0.0154	0.6182	91.6	342.3
Cylinder pressure	0.9476	0.9868	219.5	228.8
Transverse flow force amplitude	0.7199	0.9496	97.6	240.8
Transverse flow force sirection	0.0666	0.7737	154.6	234.9
Torsional vibration camshaft	0.881	0.9007	218.6	236.7
Rotational speed	0.2416	0.545	106.4	367.3

Table 5.2 Evaluation of valve plate deflection

Factor	Polynom fit		Data range	
	linear	quadratic	Min.	Max.
Guidance clearance	0.9938	0.9993	0.610	1.483
Seat ring angle	0.7009	0.7068	0.912	1.283
Modal guidance deformation	0.9102	0.9282	0.638	1.018
Valve clearance	0.0194	0.5693	0.984	1.090
Valve spring stiffness	0.5121	0.699	1.006	1.182
Valve spring misalignment	0.448	0.6672	0.704	1.075
Transverse force amplitude from elephant foot	0.0452	0.8818	1.018	2.836
Cylinder pressure	0.0983	0.3428	1.015	1.023
Transverse flow force amplitude	0.7684	0.9228	0.911	1.044
Transverse flow force direction	0.5737	0.9439	0.906	1.509
Torsional vibration camshaft	0.4055	0.9083	1.013	1.029
Rotational speed	0.5295	0.9201	0.793	1.545

as possible. The elimination of factors requires no qualitative results of the factor influence. Instead, the relative influence of each factor compared to the other factors determines the relevant input factors. Thus, a screening plan is required for this analysis. The Morris method, introduced in Sect. 2.2.2, is utilised. 12 factors require 13 simulation runs per design run. References [6, 13] suggest 4–10 design runs.

For the presented analysis the number of design runs is set to 8 which results in an overall simulation run number of 104. The number of design runs is on the upper side of the suggested range and is hence assumed to deliver decent accuracy

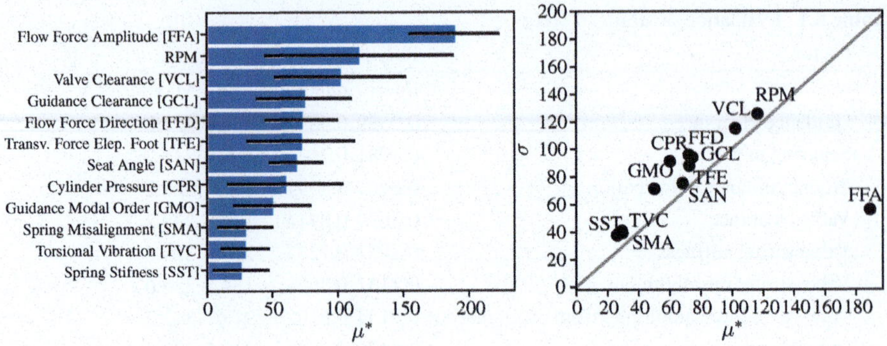

Fig. 5.5 Morris evaluation of factor influence

of results. Moreover, the number of simulation runs is kept to a reasonable level. The results are evaluated for the main output value. The evaluation determines the mean of the elementary effects and the standard deviation of the elementary effects. This delivers two main findings: The higher the mean of the elementary effects the higher the influence compared to the factor with a smaller mean of elementary effects.

Figure 5.5 shows the results of the analysis. The left graphic shows the ranking from most to least influential factors regarding their mean of elementary effect on the output value, illustrated by the blue bar in the graphic. Moreover, the standard deviation of the elementary effect is shown by the thin black bar. The factors with the least influence on the strain are the spring stiffness, the spring misalignment, the guidance modal deformation and the camshaft torsional vibration. These factors are eliminated for further analyses. The choice of 4 factors to be eliminated is a compromise. The accuracy decreases with every non-considered factor. If the factor has a small influence, the model error increase is negligible. However, the number of required simulation runs for subsequent factor influence analysis decreases significantly with a decreasing number of factors. The elimination of four factors for the case of the valve load sensitivity towards the influencing factors is assumed to hold the best balance of efficiency and accuracy.

The second finding from the Morris analysis is the kind of factor influence. This result in form of a covariance plot is shown on the right graphic of Fig. 5.5. This is derived from the ratio of standard deviation of elementary effects and the mean of the elementary effects. For a ratio smaller than 1 linear effects can be assumed. For an even smaller ratio of 0.1 or lower the factor has nearly no influence on the output value. Ratios greater than 1 indicate either a non-linear influence of the factor on the output value or high interaction effects with other factors [6, 9].

The factors have non-linear effects on the output value strain or interaction effects, respectively. Only the factor of flow force amplitude has a linear influence on the output value. The bar in the left side bar plot shows the mean of elementary effects, the additional line shows the 95% confidence interval of the resulting output value.

5.1.2 *Input Factor Influence*

The specific magnitude of influence holds important information regarding the component and system design process, because the modification of more influential parameters has a higher impact on the output value. Moreover, if the boundaries of each factor and the specific influence are known, the resulting output for every possible factor combination can be determined. By this, the maximum expectable strain of the valve can be predicted.

For the present case of the valve load output, most parameter boundaries can be clearly determined, for some boundaries reasonable assumptions have to be made. The rotational engine speed and the cylinder pressure depend on the engine operation point and are therefore known. The guidance clearance can be determined, because the manufacturing tolerances of valve and guidance are known. The variation of the flow force amplitude can be determined from measurements due to the evaluation of the cylinder pressure variance during valve opening. The valve clearance variation can be determined from measurement evaluation of the real pushrod forces (see Sect. 4.1.5.2). The transverse force from the elephant foot is assumed be considering the maximal geometric misalignment and tolerances from rocker arm to valve. However, this assumption has not been verified, yet. Another parameter with boundaries based on assumptions is the flow force direction. As mentioned at the beginning of this chapter, this variation has not been determined, yet. Therefore, the variation of the application angle of $\pm 45°$ is assumed. Hence, the main flow direction remains from cylinder middle to the cylinder wall.

By resampling various factor combinations, a distribution of strain data is produced. The higher the number of resamples, the higher the convergence of the resulting distribution. Each sample is thus one simulation run that requires resources, especially time. The generation of a meta-model helps to decrease simulation time. The meta-model is a polynomial describing the factor influence on the output value. The evaluation of a polynomial instead of a full simulation model, in the present case the OVM, is much more time efficient, so that very high numbers of resamples can be evaluated. The coefficients of the resulting polynomial also determine the sensitivity of the output towards the influencing factors.

The thesis utilises an analytical approach to determine the polynomial. For an analytical approach mostly linear or quadratic models are used as higher order models would lead to unreasonable numbers of required design runs to determine the model [15]. A DoE is chosen to generate the meta-model. The CCD is suitable for a quadratic model approach. For 8 factors the CCD leads to 324 simulation runs.

The process of generating the meta-model follows the steps introduced in [5]. The paper describes the generation of a meta-model of a turbocharger. This turbocharger meta-model is subsequently used to determine the sensitivity of the output value excitation of vibration towards constructive parameters. Hence, the overall procedure is similar to the aimed procedure in this subsection.

For the generation of the polynomial after performing all necessary simulation runs, there are two different methods introduced by [15]. Not all possible terms of

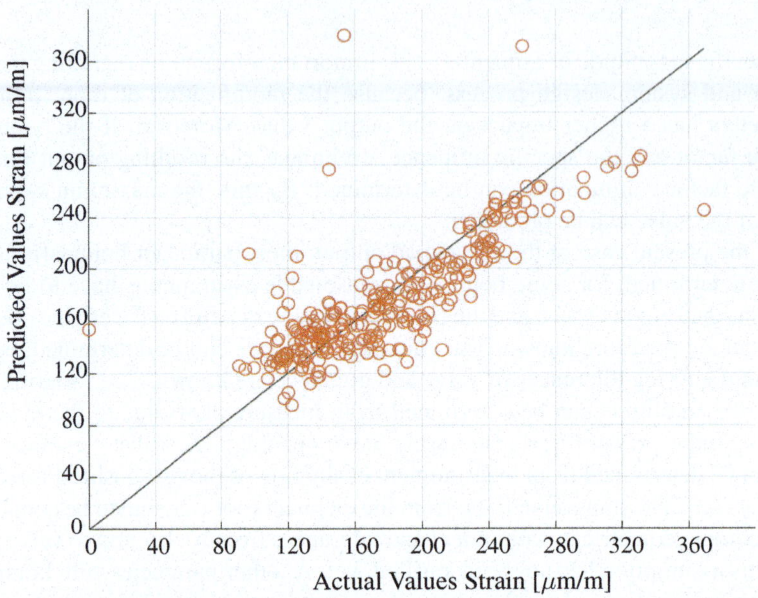

Fig. 5.6 Residual plot of regression model

the polynomial have to be used. They all require an attribute to rate the quality of the polynomial. An obvious attribute is the coefficient of determination R^2_{det} that was introduced in the previous section.

As a first step, all terms that can be part of the polynomial are defined. For the quadratic approach quadratic terms, linear terms and linear interaction terms are considered. The interaction terms are limited to interactions of two factors.

One method of selecting the terms for the polynomial is the sequential forward selection. This method starts with a simple polynomial, usually only a constant that is possibly complemented by already known terms. Iteratively, terms that are included by the full model are added to this basis polynomial. The meta-model quality is checked after each addition of a term. If the additional term improves the model quality by a certain predefined value, the term is permanently added to the polynomial.

The sequential backward selection has the opposite procedure. The start polynomial includes every term of the full model. Iteratively, a term is eliminated from the polynomial, then the model quality is checked. If the model quality improves, the term is permanently eliminated from the model. The sequential backwards method is applied to the data.

The results of the regression show high residuals (see Fig. 5.6) for higher strain values. The accordance of the simulated and the real strain data is not good enough

Table 5.3 Deviation of simulation results and regression result for overfitted regression

Evaluation Parameter	Value [%]
Mean Deviation Simulation/Regression	611.12
Max. Deviation Simulation/Regression	8818.76

to predict the factor dependency of the strain output of the simulation model. This is also indicated by the coefficient of determination at a value of $R^2_{det} = 0.6671$.

It is possible to increase the complexity of the regression by adding cubic terms, quadratic interaction terms and triple interaction terms. But a higher complexity model has the risk of resulting in an overfit of the data [15]. An overfit is the term for a model fit that describes the data used to generate the model with decent accuracy, but has high residuals by describing other data points. The transferability of the overfitted model is poor.

To improve the coefficient of determination, higher order terms are added to the meta-model. Subsequently, the newly generated meta-model is checked for an overfit. The check is based on the generation of additional test samples, which consist of factor combinations not being part of the CCD. The data overfit for the strain data is obtained by the generation of test samples. These test samples include only factor combinations that are not part of the CCD. The LHD sample used in Sect. 5.1.3 is used to test the regression. Therefore, the table includes data of 300 samples that are not included in the CCD.

The factor combinations are applied to both the OVM and the polynomial regression. Although the polynomial fits well to the CCD data with a coefficient of determination $R^2_{det} = 0.8834$, the test factor combinations lead to results that differ significantly from the simulation results. This indicates a classic overfit. Table 5.3 shows the overfit of the data. The mean deviation of simulation reference data to the data generated with the meta-model is 611 %. The highest deviation of results generated with the OVM and results generated with the meta-model is 8818.76%.

The regression results lead to the conclusion that it is not possible to predict valve strain in dependency of the relevant factors by a quadratic model. This result is a typical result for problems with a higher number of influence factors. The meta-model quality is in general decreasing with increasing number of factors [10].

Reference [5] use a linear model to describe the excitation vibration of the turbocharger, to decrease the number of time costly simulation runs. Although they also find that the model is not able to predict the resulting vibration amplitudes sufficiently, they were able to detect a jump in the vibration amplitudes. This means that the vibrations are either very small or jump to a certain, significantly higher level. This jump is caused by the excitation of sub-synchronous vibrations. If these sub-synchronous vibrations are excited, the jump in the vibration amplitudes appears. If the sub-synchronous vibrations are not excited the vibration amplitudes remain small. Therefore, the formerly continuous problem can be broken down to a discrete problem with the two possible states excitation of sub-synchronous vibrations or no excitation of sub-synchronous vibrations. Reference [5] apply the model to that finding, so that it is able to predict the appearance of the sub-synchronous vibration and therefore the appearance of high amplitudes.

This quasi-discretisation is a workaround for problems with discrete output values. The application of this workaround is not possible for the strain, as it is continuously distributed and shows no jumps or other discrete states. There are further options of meta-modelling, e.g. the determination of a meta-model by utilising neural networks. Because this approach is out of scope for the present thesis the generation of a meta-model for the load data estimation is rejected.

Nevertheless, the sensitivity of the output towards the factors can be determined. There are several methods of sensitivity analysis beside the regression approach. Some of the sensitivity approaches like the Sobol indices require a lot of additional simulation runs. Another more efficient group of methods of sensitivity determination is the group of Fourier amplitude sensitivity test (FAST). This analysis provides the first order sensitivities of the factors [13]. It is based on the analysis of the factor space by a single curve. This curve is described by the parameter s_f. The levels of a factor $x_{f,i}$ are replaced by (5.4).

$$x_{f,i} = G_{f,i}(\sin(\omega_{f,i} \cdot s_f)) \qquad i = 1, ..., n_f \qquad -\infty \leq s_f \leq \infty \tag{5.4}$$

$G_{f,i}$ are transformation functions, $\omega_{f,i}$ are frequencies that correspond to the factors $x_{f,1}, ..., x_{f,n_f}$. s_f is the control variable of the FAST. If s_f is varied, all factors vary along a curve through the factor space, too. The factors oscillate in dependency of their specific frequency $\omega_{f,i}$. The output Y_i oscillates in dependency of the factors $x_{f,i}$. For a factor with high influence on the output the amplitude spectrum of the output variable shows oscillations with higher amplitudes at the frequency corresponding to that factor than at frequencies corresponding to less influential factors. Reference [14] gives further information on the choice of $G_{f,i}$ and $\omega_{f,i}$. There are different types of FAST requiring different kinds of samples for the analysis. The first FAST analysis presented here is the random balanced design (RBD) using a LHD based sample for the analysis. In contrast to this, the other FAST method used in this thesis is the extended Fourier amplitude sensitivity test (eFAST) method which requires a specific sample for the sensitivity analysis.

The RBD adaptation within the FAST method reduces the number of frequencies which leads to a curve that is not space-filling [17]. To generate a space-filling design, permutations of points describing that curve are utilized. In that context the LHD provides a suitable space-filling design to be used for this sensitivity analysis method [8]. The analysis in Sect. 5.1.3 also requires a LHD sample. Hence, no additional simulation runs are necessary for the determination of the first order indices. The resulting sensitivity indices are shown in Fig. 5.7. The sum of the first order effects is approx. 50% of the overall effects on the output value. That means that in turn 50% of the overall output is a result of interaction effects. The dependency of the strain from the relevant factors is hence complex, the interaction of factors is not negligible. The flow force amplitude (FFA), the engine speed (RPM) and the valve clearance have the biggest influence regarding the first order effects.

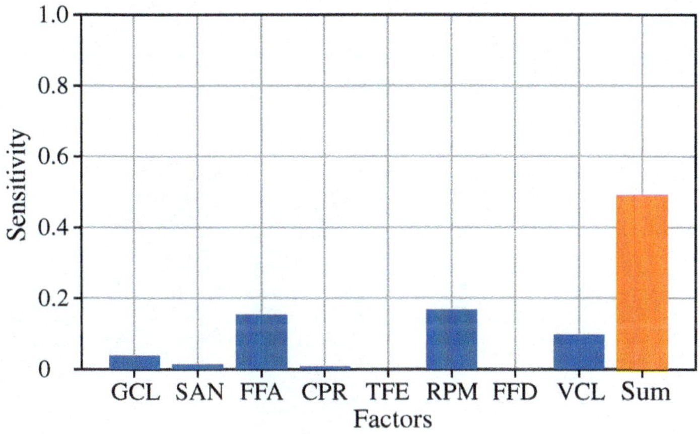

Fig. 5.7 First order sensitivities of valve strain

As mentioned above, the standard FAST method as well as the RBD-FAST method deliver the first order sensitivity indices. These describe the main effects of the factors. The sum of all first order indices is therefore 1 at a maximum.

There are furthermore the total order indices that include all interaction effects of the respective factor. Therefore, the sum of the total indices can be higher than one as the factors can have first order effects and additional interaction effects that are ascribed to all factors contributing to a certain interaction effect. To determine these total indices the eFAST method is utilized. This method requires a specific data sample. The LHD sample can not be utilized. The size of the required data sample depends on the number of factors and the order M_{FAST}. This order M_{FAST} determines how many harmonics are considered. Reference [17] specifies the usual order of $M_{FAST} = 6$, [14] suggest an order of $M_{FAST} = 4$ or higher. With that minimum order the minimum eFAST sample of 8 factors is 520. With an assumed simulation time of 2.5 h and 2 parallel simulation runs this would lead to 650 h or almost 4 weeks of simulation. To decrease the effort the analysis is limited to the three factors with the highest first order effects. This analysis requires 120 simulation runs. Figure 5.8 shows the results of this analysis. The sensitivity results of the three most influencing factors show that the rotational speed influences almost 80 % of the output but only approx. 25% of the influence is caused by the main effect of this factor. Also the flow force amplitude has a high influence on the output. The total index shows an influence of approx. 75% on the overall model output. The dominant ratio is hence the interaction effects of the factors.

This sensitivity analysis results lead to three main conclusions:

1. The influence of the factors on the output valve strain is complex and can not be described by a quadratic model.

Fig. 5.8 Sensitivity indices extended FAST analysis

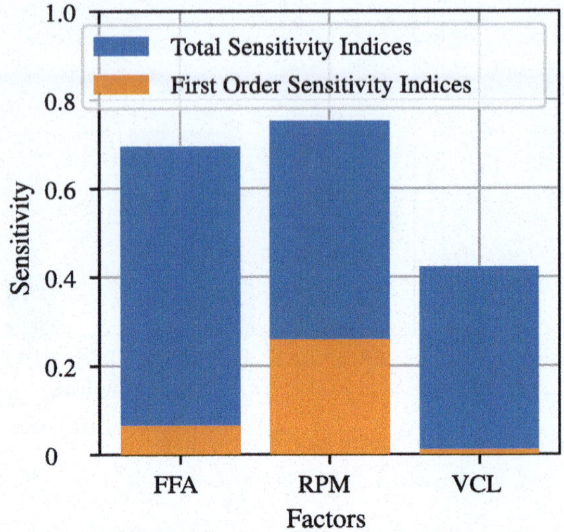

2. The most influencing factors are the flow force amplitude, the engine rotational speed and the valve clearance. However, the dominant effect on the output value strain is caused by interaction effects.
3. The determination of a meta-model can save a lot of simulation time. Furthermore, the detailed investigation of the interaction effects is possible. However, a more extensive approach of meta-model generation is necessary.

5.1.3 Determination of the Valve Load Collective

The improvement in design process is a side benefit of the determination of factor influence. The main aim is the simulation based determination of the load collective for the fatigue strength assessment. This load collective distribution of the resulting strain is generated by a variation of the influencing factors within their previously defined limits. Due to the complex factor influence, the generation of a meta-model was not successful. Hence, the variation of influencing factors in order to obtain a valve load collective has to be performed for the full model, the OVM.

For this purpose the distribution of each factor has to be known. There are two options to determine the distribution of the factor values within these limits. The first option is to determine the distribution based on measurement. The second option is to assume a distribution.

Because there is no data on the factor distributions, these distributions have to be assumed. Reference [7] assume a normal distribution of the factors influencing the fatigue strength of bolted connections. A plausible approach on the normal distri-

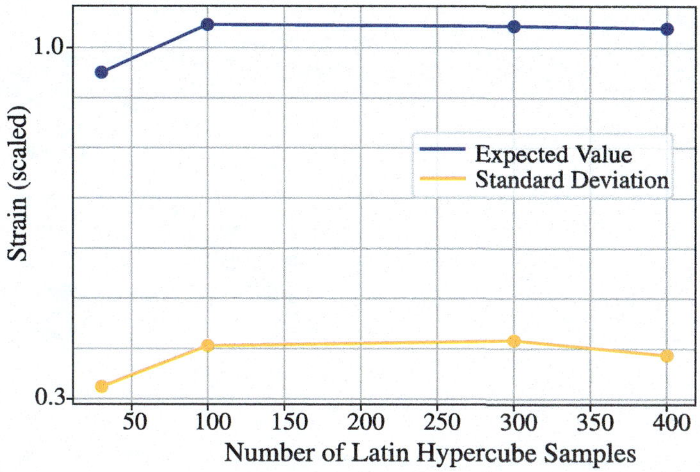

Fig. 5.9 Convergence of expected value and standard deviation for LHD samples

bution assumption takes the nominal factor values as the mean. The variance of the factors has to be assumed. The normal distribution leads to less factor combinations with factor values close to their specific limits as the factors concentrate more likely around their specific nominal values. This is realistic, as the normal distribution is likely to be a sufficient assumption of the factor value distribution, because of the CLT introduced in Sect. 2.2.1.1. The assumption provides the risk of an underestimation of the frequency of factor value appearance close to the their limits. Therefore, this thesis assumes a random distribution of all factor values within their limits to decrease the risk of underestimating extreme factor combinations. This is a more conservative assumption.

It is now important to choose a space-filling design in order to get a complete field of potentially occurring strain data. For a random based distribution the Monte Carlo sampling and the Latin Hypercube sampling provide space-filling designs. [16] compare both approaches and show that the mean value of the analysed output converges significantly faster by using the LHD. Hence, this design decreases the number of required samples.

As a consequence, the generation of a load collective utilizes a LHD with random distributed factor values. The LHD has no requirements regarding minimum sample size, but the resulting data are not convergent for too small sample sizes. A study on the number of samples ensures the convergence of the output. Different numbers of samples are evaluated regarding their convergence.

The load data collective determination is repeated for 30, 100, 300 and 400 samples. Mean and standard deviation of the generated strain data show a good convergence for 300 samples. Figure 5.9 shows the convergence results of the test samples.

The mean value of the data is constant from 100 samples. The standard deviation increases slightly between 100 samples and 300 samples and decreases slightly at

Fig. 5.10 LHD with good and bad quality, [15]

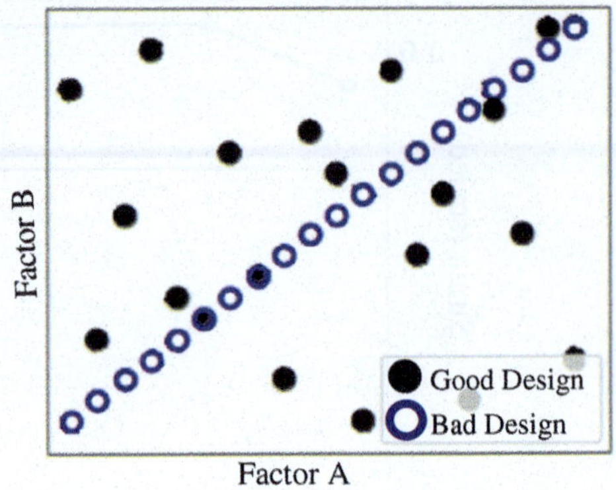

400 samples. This might be a result of variable quality of LHD designs: The quality of the space-filling character of the design is variable for constant sample sizes because the recombination of permutations are random. This problem is addressed in [15] and shown in Fig. 5.10. Hence, the slight variation of the standard deviation might be a result of varying quality of the designs.

The quasi-constant values of mean and standard deviation are reached for 100 samples. To generate security towards the varying quality of the LHD and possibly higher variances of factors for other engines the suggested number of samples is 300. The load collective is supposed to include the variance of the strain load for all cylinders of all engines of the same type during the required operation hours of the valve.

To verify the simulated load collective, this load collective is compared to the load collectives of the exhaust valves that are determined by the measurements presented in Sect. 4.1. This data only represents a load collective of one cylinder of one specific engine. Therefore, the variance of the simulated load collective is supposed to be slightly higher. Nevertheless, this comparison shows whether the magnitude of the predicted load data are realistic and relate to the measurement data. Figure 5.11 shows the comparison of the load collectives. All strain data are scaled by the same factor to ensure comparability. The data of both exhaust valves in the measurements differ. Exhaust valve 1 has an overall higher load than exhaust valve 2. Therefore, valve 1 builds the main basis for this comparison. The magnitude of the strain amplitudes is in a similar range. However, for the simulated load collective high strain amplitudes appear in a slightly higher frequency, therefore the simulated strain data shows a higher slope in the right part of the plot. Thus, the simulated load collective is more conservative than the measurement data. This is a desired result as the simulative load collective is supposed to include a higher variance of the load data. Nevertheless, the maximum strain load is very similar, so that the simulated load collective is not too conservative.

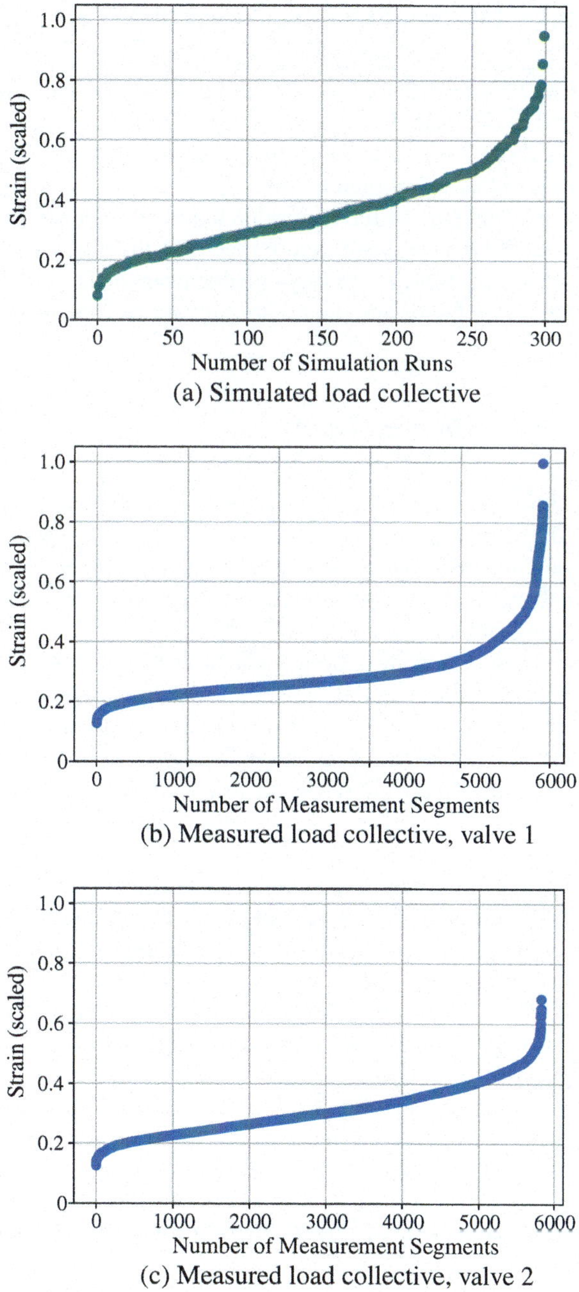

Fig. 5.11 Comparison of strain data collectives

(a) Simulated load collective

(b) Measured load collective, valve 1

(c) Measured load collective, valve 2

5.2 Fatigue Strength Assessment

The fatigue strength assessment relates the actual component load to the maximum endurable load. The fatigue strength assessment presented in the following uses the local stresses to determine the actual loads of the component, following the approach in [11] that is outlined in Sect. 2.3.

On the side of the endurable loads, the relevant material properties of the valve steel X45CrSi9-3 are determined. On the side of the actual loads, the equivalent strain and stress amplitudes are determined. The result of the fatigue assessment are the safety factor and the failure probability, respectively.

5.2.1 *Material Data*

This section examines the side of endurable loads of the fatigue strength assessment. The valve steel usually chosen for the MAN ES valves is not included in the material data provided by the FKM guideline. The valve steel norm [3] gives minimum values for the tensile strength and the yield strength for room temperature and for temperatures of more than 500 °C.

This is not an optimal data base for several reasons. With the data given by the norm the material data has to be interpolated for temperatures between room temperature and 500 °C. This is a rough assumption, because most strength values of steel do not behave linearly in dependency of the temperature. Moreover, the norm gives the lower limit of the specific value. The real valve steel is likely to have higher strength values. Therefore, the data from this norm can lead to conservative safety factors in the fatigue assessment. As the norm provides the lower limit, the provided strength data can be assumed to fulfill the 97.5% survival probability required by [11]. Although the FKM guideline provides equations to determine material data of fatigue limit from tensile and yield strength, the survival probability can differ for the fatigue limit data.

The most reliable material data is data obtained from fatigue test with specimen of the actual steel charges that are used for the valve manufacturing. The investigations for this thesis have no access to material data from material tests. Instead, this thesis uses data provided by the data sheet from Deutsche Edelstahlwerke [2]. This data sheet includes S-N curves of the valve steel for higher temperatures at a survival probability of 50%. Figure 5.12 shows the S-N curve provided by the data sheet.

The material data $S_{X,50\%}$ with a survival probability $P_S = 50\%$ are converted to material data $S_{X,97.5\%}$ with a 97.5% survival probability according to [11]:

$$S_{X,97.5\%} = \frac{S_{X,50\%}}{j_{n,S}} \tag{5.5}$$

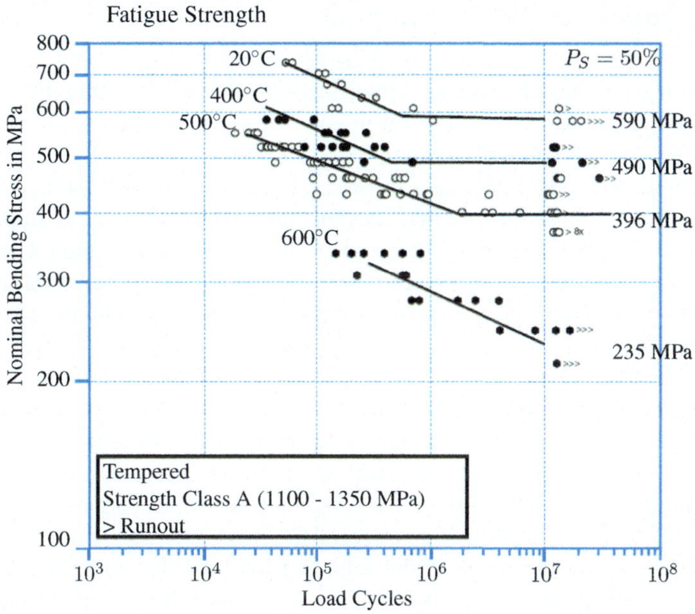

Fig. 5.12 Fatigue strength of X45CrSi9-3, [2]

$j_{n,S}$ is the statistical conversion factor that is determined by the number of individual tests n and the logarithmic standard deviation LSD that is an empirical value given by the FKM guideline. For the fatigue strength of non-welded components the LSD equals 0.04. From Fig. 5.12 a minimum test number of $n = 16$ is determined. The statistical conversion factor is defined as:

$$j_{n,S} = 10^{(2+1/\sqrt{n})\cdot LSD} \tag{5.6}$$

The conversion from $P_S = 50\%$ to $P_S = 97.5\%$ means a parallel shift of the S-N curve in negative ordinate direction.

The temperature of the radius region is between 400 °C and 500 °C. Therefore, the data is linearly interpolated for the radius temperature. The assumption of a linear behaviour of the material data between these temperatures is acceptable, as the S-N curves lie close to each other, so that the linear characteristic is only assumed for a small range of the material data. With the material data from Fig. 5.12, the statistical conversion factor (5.6) and the interpolation of the material data the slope k_{mat} of the S-N curve before the knee point and the number of cycles at the knee point N_D of the S-N curve can be determined.

5.2.2 Equivalent Strain Amplitude

This section examines the side of actual loads of the fatigue strength assessment. The exhaust valves have a predefined number of operation hours after which they are replaced. These operation hours range from 30000 to 40000 h.

The fatigue strength assessment follows the approach of the damage-equivalent amplitude that is introduced in Sect. 2.3. The iteration process of the amplitude follows the Miner consistent hypothesis and uses the S-N curve data k_{mat} and N_D determined in the previous subsection. Moreover, the iteration uses the number of operation hours and the load spectrum which is determined by simulation as explained in Sect. 5.1.3 as an input.

Although [11] explains the damage-equivalent amplitude in the context of stresses, the procedure is dimensionless. It is therefore applied directly to the strain data of the load spectrum and delivers a damage-equivalent strain amplitude. The load-spectrum as well as the damage-equivalent amplitude must not be considered as absolute load values of the valve. They rather provide a comparison value. This is important, because the sensor position of the model the load collective is based on is not necessarily the plane of the highest total load. However, the load of the valve during closing is almost pure bending. Therefore, it is assumed that close to the exact position of maximum valve load, the load at the sensor position behaves linearly proportional to the stress at the exact position of maximum load. Therefore, the load case resulting in maximum strain in the simulation is also the load case that results in the maximum load on the real valve.

The determination of the damage-equivalent strain amplitude is therefore not directly used for the fatigue assessment. Instead, the simulation parameter of the OVM are adjusted in a way that the resulting strain maximum is the damage-equivalent amplitude. In this context, the simulation of the valve working cycle is repeated in order to obtain the distribution of stresses and strains for the whole valve body. SIMPACK generates stress data and strain data. The modal stresses consisting of the modal coordinates and the mode shapes are handed over to FEMFat via the FEMFat *ChannelMax* module. Within this module each eigenmode stress is defined as a separate channel. Furthermore, each inertia relief mode is defined as a separate channel. The total number of channels depends therefore on the number of eigenmodes and inertia relief modes defined during the flexible body reduction process. FEMFat determines the maximum load from these channel informations by superpositioning the channel information of all available channels. FEMFat uses the original FE geometry including the mesh to determine the individual load of each node from the "ChannelMax" load data. Additionally, fatigue related influences and material data are applied to individual nodes or node groups.

The temperature distribution from the temperature measurement evaluation with Ansys Workbench module *Steady-State Thermal*, shown in Sect. 4.1.5.1, provides the temperature data of the nodes. The temperature influence is considered following the FKM guideline and shown in Sect. 2.3. The equivalent stress hypothesis selected for

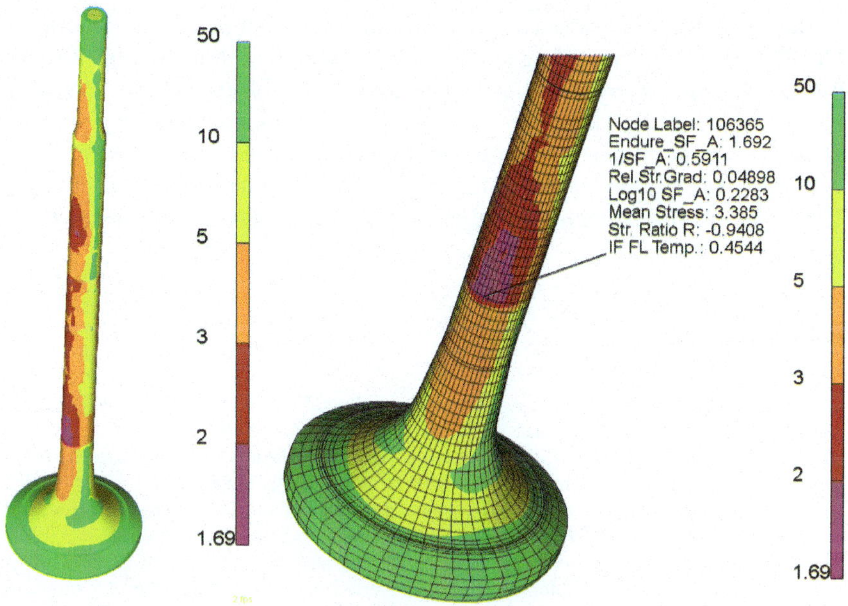

(a) Overview over safety factor distribution (b) Minimum safety factor of exhaust valve, visible valve mesh

Fig. 5.13 Fatigue assessment results of the 49/60 exhaust valve

this analysis is the scaled normal stress in critical plane hypothesis [4]. The resulting minimum safety factor is the reciprocal of the cyclic degree of utilization.

The minimal safety factor S_F of the 49/60 exhaust valve regarding fatigue strength is 1.69. The required safety factor according to MAN ES guidelines is 1.5. Thus the exhaust valve fulfils this requirement. The location of minimum safety factor is above the valve plate radius and therefore slightly higher than the sensor position. The position is specific to that valve geometry because the valve stem has a conic expansion at the bottom of the stem. Other valve geometries have a stem that directly transitions into the valve plate radius. But this position is in accordance with findings from other publications dealing with valve fatigue (see Sect. 3.2). The temperature in this area is higher than the temperature at the valve radius, which indicates the plausibility of this position to be the one with the lowest safety factor.

The assumption of an almost pure bending load of the exhaust valve during closing is verified by the FEMFat evaluation, shown in Fig. 5.13, because the results show a mean stress of <4 MPa.

The degree of utilization is approx. 60%. The determination of the failure probability assumes the load to be a safe load assumption following the definition of FKM. There are empirical values of material strength scattering. This empirical data is well evaluated and has a good accordance with real material behaviour. The scattering is defined by the already introduced variable *LSD*.

Beside the standard deviation, the form of the distribution has to be determined. The FKM guideline according to [11] assumes the logarithmic normal distribution. By applying this distribution the failure probability for the 49/60 valve results to $P_{A,log} \approx 3.5 \cdot 10^{-15}$. The more conservative distribution is the normal distribution. With the normal distribution the failure probability of the exhaust valve is $P_{A,N} \approx 2.7 \cdot 10^{-8}$. The required maximal failure probability is one part per one million parts. The requirement relating to the failure probability is therefore fulfilled, too.

References

1. Dassault Systèmes Simulia Corp.: *Simpack - SIMULIA User Assistance 2022x*. 2021
2. Deutsche Edelstahlwerke: *Werkstoffdatenblatt: X45CrSi9-3*. 2016
3. DIN Deutsches Institut für Normung e.V.: *Ventilstähle und -legierungen für Verbrennungskraftmaschinen: Deutsche Fassung EN 10090 : 1998*. Berlin, 1998
4. Gaier, C. ; Hofwimmer, K. ; Dannbauerr, H. : Accurate and Efficient Methods for Multiaxial Fatigue Analysis. In: NAFEMS (ed.): *NAFEMS World Congress 2013*, 2013
5. Göbel, S. ; Daniel, C. ; Woschke, E. ; Strackeljan, J. : DoE basierte Sensitivitätsanalyse konstruktiver Lagerparameter eines gleitgelagerten Abgasturboladers. In: *11. Internationale Tagung Schwingungen in rotierenden Maschinen* (2015)
6. Iooss, B. ; Lemaître, P. : A review on global sensitivity analysis methods. In: Dellino, G. (ed.) ; Meloni, C. (ed.): *Uncertainty Management in Simulation-Optimization of Complex Systems*. Springer US, 2015
7. Lehmann, M. ; Hummel, D. : Fatigue verication of high loaded bolts of a rocket combustion chamber. In: *11. Weimarer Optimierungs- und Stochastiktage*, 2014
8. McKay, M. D. ; Beckman, R. J. ; Conover, W. J.: Comparison of Three Methods for Selecting Values of Input Variables in the Analysis of Output from a Computer Code. In: *Technometrics* 21 (1979), Nr. 2, 239–245.
9. Morris, M. D.: Factorial Sampling Plans for Preliminary Computational Experiments. In: *Technometrics* 33 (1991), Nr. 2, 161–174
10. Most, T. ; Will, J. : Robust Design Optimization in industrial virtual product development. In: *Proceedings 5th International Conference on Reliable Engineering Computing, Brno*, 2012
11. Rennert, R. ; Kullig, E. ; Vormwald, M. ; Esderts, A. ; Siegele, D. : *Rechnerischer Festigkeitsnachweis für Maschinenbauteile: Aus Stahl, Eisenguss- und Aluminiumwerkstoffen*. 6. Frankfurt (Main) : VDMA, 2012
12. Rieß, A. : *Model order reduction based simulation and optimization of large bore internal combustion engines*. Universität Augsburg, Mathematisch-Naturwissenschaftlichen Fakultät, Dissertation, 2014
13. Saltelli, A. ; Tarantola, S. ; Campolongo, F. ; Ratto, M. : *Sensitivity analysis in practice: A guide to assessing scientific models*. Repr. Chichester : John Wiley & Sons, 2007
14. Saltelli, A. ; Tarantola, S. ; Chan, K. P. S.: A Quantitative Model-Independent Method for Global Sensitivity Analysis of Model Output. In: *Technometrics* 41 (1999), Nr. 1, 39-56.
15. Siebertz, K. ; Bebber, D. van ; Hochkirchen, T. : *Statistische Versuchsplanung*. Berlin, Heidelberg : Springer Berlin Heidelberg, 2017
16. Stockinger, A. ; Wartzack, S. : Application of Sampling Methods for Statistical Tolerance Analysis. In: *7. Weimarer Optimierungs- und Stochastiktage*, 2010
17. Tarantola, S. ; Gatelli, D. ; Mara, T. A.: Random balance designs for the estimation of first order global sensitivity indices. In: *Reliability Engineering & System Safety* 91 (2006), Nr. 6, 717–727.

Chapter 6
Conclusion and Outlook

This thesis had the aim of generating a method to assess the fatigue strength of exhaust valves of heavy-duty engines predictively and efficiently. For the determination of the loads needed for the fatigue strength assessment, the valve dynamic had to be determined first.

Simulation models according to the current state of the art are not able to reproduce the dynamic of the heavy-duty engine valves. This thesis determined the dynamic flow forces resulting from a pressure gradient in the cylinder as the main cause for the valve secondary dynamic. The implementation of these dynamic forces to an MBS model of the valve train complements the known influences on the valve train dynamic regarding the state of the art. The development of a simplified, predictive method to determine resulting flow forces on the valve is an outlook on future improvement. Currently, the CFD based determination of flow forces is costly regarding simulation effort.

This thesis mainly used the OVM to simulate the valve dynamic. This is a simplification compared to other valve train models with similar purpose. The OVM provides the benefit of a high modelling depth with focus on the valve deformations and loads by keeping the simulation effort as low as possible. The validation of the model by measurement data proved that this simplification is sufficient regarding valve dynamic and valve loads for a known valve clearance. However, a comparison between the CVT and the OVM for varying valve clearances and valve lifts can improve the boundary conditions of the OVM.

The analysis of the valve dynamic showed a high interaction of the valve with the surrounding system. The valve dynamic is influenced by the flow forces. Hence, changes in thermodynamic boundary conditions lead to a change of valve loads. A change in the valve design on the other hand leads to changes in the exhaust air stream and thus influences the quality of the engine's firing process. Therefore, an overall system analysis is necessary, including thermodynamical and mechanical boundary conditions, to increase the quality and efficiency of the design process.

A. Eret, *Dynamic and Fatigue Assessment of Heavy-Duty Engine Valves*, Mechanics and Adaptronics, https://doi.org/10.1007/978-3-031-49151-1_6

The method presented in this thesis and the provided model deliver one component of this systemic analysis of the valve train.

The simulation model validation by measurements on a fired engine is a further complement to the current state of the art. Usually, valve train measurements are carried out on trailed engines. The measurement setup used in this thesis is much closer to the real operation conditions of the serial engine. As a result, the quality of the simulation validation is higher for the fired engine. The direct application of strain gauges to the valve is a further important aspect of the measurement setup. It enables a quantitative validation of the actual valve loads.

The statistical evaluation of the measurement data added findings on the distribution of valve load data and builds the basis for the determination of the valve load for the fatigue strength assessment. Furthermore, the implementation of DOE methods to determine the valve load collective increases the reliability of the fatigue strength assessment, because variations of load influencing factors are considered. However, the influence of factors on the valve load is complex. Several simulation runs are necessary to determine a sufficient load collective. A meta-model reduces the simulation effort significantly. For the case of valve load, the determination of such a meta-model requires advanced algorithms, e.g. neural networks. A further improvement of the valve load collective determination is the determination of the real distribution of each influencing factor. The presented methods assumes a random distribution. The consideration of the real distribution of the influencing factors increases the quality of the load collective and reduces conservativeness in the generation of the valve load collective. The result is a more efficient valve design.

This thesis provides a predictive and transferable method to determine the fatigue strength of heavy-duty engine valves considering the valve dynamic. The implementation of this method allows a systemic consideration of the valve design in its mechanical and thermodynamical environment. The fast adaption of variable geometries and boundary conditions to the valve design evaluation is the main benefit of this method.

Appendix A

A.1 Process of Valve Fatigue Assessment

See Fig. A.1.

A. Eret, *Dynamic and Fatigue Assessment of Heavy-Duty Engine Valves*, Mechanics and Adaptronics, https://doi.org/10.1007/978-3-031-49151-1

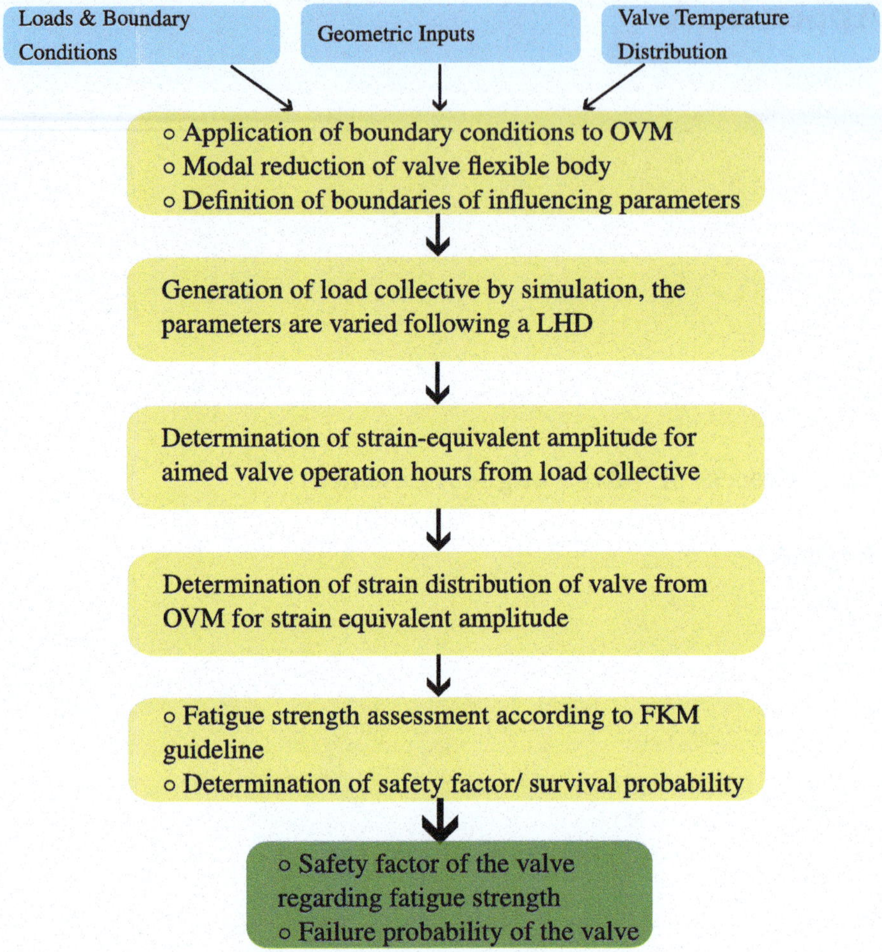

Fig. A.1 Flow diagram of valve fatigue assessment method

The manufacturer's authorised representative in the EU is Springer Nature Customer Service Centre GmbH, Europaplatz 3, 69115 Heidelberg, Germany. If you have any concerns regarding our products, please contact ProductSafety@springernature.com

Printed and bound by CPI Group (UK) Ltd, Croydon, CR0 4YY
28/09/2026
02224078-0010